HIMALAYA

Published in cooperation with the Center for American Places,
Santa Fe, New Mexico, and Harrisonburg, Virginia

Life on the Edge of the World

HIMALAYA

DAVID ZURICK AND P. P. KARAN

MAPS BY JULSUN PACHECO

THE JOHNS HOPKINS UNIVERSITY PRESS
BALTIMORE AND LONDON

FRONTISPIECE: *Mt. Ama Dablam provides a spectacular backdrop to the Khumbu region of Nepal. Some of the highest mountains in the world are located here, including Mount Everest. The people who live in this region, known as the Sherpa, cultivate potato fields and keep herds of sheep and yak. Photo by P. P. Karan*

© 1999 The Johns Hopkins University Press
Maps © 1999 Julsun Pacheco
All rights reserved. Published 1999
Printed in the United States of America on acid-free paper
9 8 7 6 5 4 3 2 1

The Johns Hopkins University Press
2715 North Charles Street
Baltimore, Maryland 21218-4363
www.press.jhu.edu

Library of Congress Cataloging-in-Publication Data will be found
at the end of this book.
A catalog record for this book is available from the British Library.

ISBN 0-8018-6168-3

Contents

Preface

IN THE SPRING OF 1996, a violent storm swept unexpectedly over Mount Everest, killing most members of a climbing party attempting to reach the summit. The group included American, European, Indian, and Japanese climbers, as well as their Sherpa guides. A journalist and a film crew were on the mountain at the time. They recorded the event for a mesmerized world. A great controversy ensued about the role of guided climbs in the Himalaya, whether it was safe or appropriate to conduct them and whether people acted badly or not in the midst of great human tragedy. A slew of books and magazine articles came out in the Western presses. A film was produced. All eyes were on the Himalaya. It was an outpouring of public interest the likes of which had not been seen since Mount Everest was first climbed in 1953 by Edmund Hillary and Tenzing Norgay.

The world's fascination with the Himalaya, spurred on by the 1996 tragedy, has gradually assumed a more comprehensive nature. People want to know more about the region, not just about mountain climbing on Mount Everest but about life amid the entire span of highlands that stretches across Asia. It is the world's most spectacular mountain range. There is a genuine concern among many quarters that life in the Himalaya is changing too fast, that the landscapes are being degraded, and that the future of the mountains is rather bleak. These impressions are mostly gained from another literature, written not by mountaineers about the high summits but by journalists and scholars about the deep valleys and inhabited ridge lines.

During the past two decades, many people have studied the environmental dilemmas that face the Himalaya. Their work falls into two distinct paradigms. The first group of studies is largely based on impressions and observations of the environment in selected villages or in small areas, mostly in central Nepal and in other accessible areas such as the Doon Valley and Nainital in India, as well as along the Karakoram Highway in Pakistan. It has led writers to generalize about a crisis in the Himalayan region. There have been reports of alarming environmental destruction in the mountains: rapid population growth; lands degraded by deforestation, scarred by landslides, and cut by quarries; streams choked with debris; and fields lost to floods and landslides. These are scenes of disaster. In this paradigm, human activities are thought to play the major role in the degradation of the Himalayan environment.

A second group of studies, also based mainly on research in Nepal, has questioned the accuracy, extent, and impact of the above model of human activities, particularly regarding forest clearing. They point to a haze of uncertainty that exists in the reports, question the accuracy of the deforestation figures, and decry the notion that forests act like sponges to soak up and store water, as suggested by scholars who see the forests to be the key to environmental stability in the mountains. This second group of studies argues that the environmental problems of the Himalaya have been exaggerated and overdramatized.

According to this second paradigm, much of what is written about the land degradation in the Himalaya is unproven and the arguments that link forest loss in the mountains to disastrous floods in India, Pakistan, and Bangladesh are tenuous. Natural processes in the Himalaya, according to this paradigm, dwarf the tiny impacts of human activities. This paradigm asserts that "the contribution of human interventions over the past three or four decades, or even centuries, has been insignificant when balanced against the natural processes at work."[1] The bulk of the sediment load in most Himalayan rivers, according to this view, comes from huge landslides which result from the rapid tectonic uplift of the mountains, the seismic activity, and the rapid incision of the rivers, and not from the land-clearing efforts of villagers.

The debunking of the Himalayan "crisis"—showing that it is a myth, an erroneous perception of human-induced land degradation perpetrated by bad information—has become a cause célèbre among some scholars. Numerous workshops and conferences have been held in its honor. A plethora of articles and books have been written about it. But the real problem, from our perspective, stems from the fact that there never was a single, unified model of environmental degradation in the Himalaya. Instead, there were numerous, discrete observations and reports about varying landscape conditions. These were compiled into a crisis of Himalayan magnitude. Once assembled, the various accounts gave the impression of a regional condition that never really existed. Much of the current confusion about environmental conditions in the Himalaya may stem from that assemblage.

But the matter is more than a mere academic curiosity or debate. The circumstances of life in the Himalaya are, indeed, changing rapidly, and with serious consequences for both the natural environment and human society. The mountains contain extraordinary biological diversity. The eastern sector, deeply imperiled by recent landscape change, has been designated a world-class biodiversity "hotspot." Fifty million people live in the mountains and gain their livelihoods from them. Several hundred million more live within sight of the Himalaya, or among places that are inextricably tied to them in ecological ways. Much of the rest of the world also has stake in the Himalaya, for it is a place of truly global imagination.

Many of the reports produced in both the paradigms are based on first impressions, or on hit-and-run microstudies conducted mainly by Western scholars, rather than being the result of careful comprehensive field observations over a long period of time, and spread across the vast expanse of the Himalaya from the Indus to the Brahmaputra rivers. The major intellectual problem with both paradigms is that the mountains are so vast and complex as to make any generalization untenable.

This book does not fall into either of the two paradigms on Himalayan environmental degradation. The studies reported here reveal major regional variations in the patterns of nature and society change, and show the importance of both human intervention and natural processes in the transformation of the Himalayan worlds. The importance of specific geographical factors varies from place to place, depending upon local natural and cultural histories. Scholars espousing one or the other paradigm seek a simplistic explanation of complex phenomena, and ignore the fact that one of the most compelling characteristics of the Himalayan region is its immense variability, which is readily apparent to even the most casual visitor. Morever, the resiliency of mountain people in the face of sometimes alarming circumstances is truly amazing. It is its own story. To see the Himalaya not just as a physical place, but as a place in the minds and spirits of its residents is to make the changes that occur there more comprehensible.

Much of the research reported here was supported by the United States National Science Foundation. The grants allowed numerous field studies which built upon our long-term association with the mountains. This book is the direct product of those studies. We examine here the history of nature and society change in the Himalaya, according to archival and census data on population, forest cover, and land-use change during the past century, in more than one hundred districts stretching from the Indus River to the Brahmaputra River. That comprehensive regional analysis is supplemented by intensive local studies conducted at seven field sites in the Kulu and Sutlej valleys (Himachal Pradesh), the Alaknanda Valley (Uttarkhand), the Annapurna Nature Preserve and the middle-mountains region (Nepal), the Richu Khola watershed (Sikkim), and the Tongsa and Mongar valleys (Bhutan). Additional field visits were made to remote sites in Zanzkar, Garhwal, and Arunachal Pradesh, including the Ziro Valley. In total, we have spent nearly fifty years conducting field studies in the Himalaya, spreading across the range from Kashmir to the Assam Valley. The seven areas noted above were selected specifically to assess the role of natural processes and development activities such as roads, agricultural projects, tourism, and rapid population growth, in producing environmental change.

Since the early 1950s, we have visited most Himalayan districts in Pakistan, India, Nepal, and Bhutan. We have traveled and lived in all the major geographical provinces of the range, over extended periods of time, observing the human, economic, social, and environmental transformations. On several of these research trips, we were accompanied by the distinguished American cultural geographer Cotton Mather and the celebrated Japanese cultural anthropologists Shigeru Iijima and Hiroshi Ishii. They have helped us to understand the complex nature/society relationships in the Himalaya. For several years we also held month-long summer field seminars in the Himalaya, in which our graduate students and well-known Indian scientists and geographers specializing in the mountains, such as A. B. Mukerji, Harjit Singh, and B. K. Roy, participated and provided fresh insights into the social and environmental transformations taking place in the area. Ongoing and fruitful collaboration with many native and Western Himalayan scholars have helped us to keep in balance the widely divergent views on changes in the region.

Our research over the years has resulted in seven books and more than fifty scientific papers since the 1960s. We have also reported our studies at numerous international conferences and workshops in the Himalayan region. Our personal experiences living and working in the mountains over a lengthy period strengthen the analysis of the archival data and the interpretation of regional environmental trends reported in this book. As we enter into a new millennium, the prospects of land and life in the Himalaya become ever more complex as the modern world encroaches upon the mountains. Because the Himalaya plays such a prominent role in Asian affairs, and infuses the imaginations of people everywhere, it is well worth mapping the course of its change.

The work reported here is unusually salient in the debate about environmental stability in the Himalaya because it brings the contemporary patterns of land degradation into a historical and social context. Thus far, few studies have attempted to merge regional and local data into a single analytical framework. The attempt to analyze in quantitative terms the Himalayan region, spanning four countries, over a hundred-year period, was a mammoth undertaking. Fortunately, this phase of research went smoothly, thanks to the valuable assistance of a number of local officials and scholars in each country whom we have come to know during our earlier fieldwork and residence in the Himalaya.

The regional analysis of the century-old data set contributes to a meso-scale scholarship of the sort needed in current research on the human dimensions of global environmental change. Although we are currently experiencing a resurgence of public interest in environmental issues, the contribution of the social sciences has not been as prominent as it might be. Environmental problems are not just of a technical nature produced by "natural processes." The persistence of environmental crises around the world is raising important social, political, and economic questions. The social sciences are called upon to make a contribution to the analysis of this important field. It is a substantial challenge. More than perhaps any other issue, the "environment" calls upon the social sciences, particularly geography, to develop synthetic approaches combining both human and natural perspectives in the analytical framework. This book is dedicated to the advancement of social-science research on the environment, to the understanding of the plight of mountain people, and to the education of all persons seeking to know more about the Himalaya. It is written not only for the academic community, from which much of it derives, but also for anyone who has the imagination to consider the mountains of southern Asia as a part of their own world.

Acknowledgments

THE STUDY reported here results from decades of fruitful collaboration with people and organizations located throughout the world. It is impossible to name them all. The major research and government agencies in Pakistan, India, Nepal, and Bhutan facilitated our visits and research in those countries. We are deeply indebted to them and to the many officials and independent scholars associated with those institutions who gave generously of their time and resources to assist our study. Without their help, our research would not have been possible.

Several individuals deserve special mention, for they contributed in substantial ways to our Himalayan studies. Dr. M. P. Thakore assisted immeasurably in the acquisition of government documents in India and always provided a welcome refuge in his home. Mr. Madan Sharma, railway guard on the tiny Shimla railroad and student of world history, informed us about the colorful history of the British hill stations. Mr. Prith Pal Singh, orchardist in the Sutlej Valley, taught us about apple growing in the Himalaya and shared with us the fruits of his labors. Mr. Sud, proprietor of Maria Brothers Bookstore in Shimla, provided rare old documents and prints for our study. A. K. Narayan showed us the way to the glaciers hanging above the Baspa Valley, where he was conducting geological studies relating to Himalayan climate change. Praveen Grover drove over 1,500 kilometers on bad roads in his jeep so that we could see some of Spiti and Zanzkar.

Several of our graduate and undergraduate students assisted in various stages of the work. Tsering Tsomo helped with archival collections, including Indian census materials and Tibetan documents. Chip Costello and Brent Winters worked on the compilation of data sets. Bilal Butt produced some of the tables. Susan Stratford helped with the preliminary maps. The graduate students in a semester seminar on the Himalaya conducted by David Zurick at the University of Hawaii provided some wonderful insights and discussions on a range of topics.

Major funding for our field studies in the Himalaya has come from the National Science Foundation (grant #SBR-9300017), the National Geographic Society, the United Nations, and the American Philosophical Society. We thank those agencies for their support of our work. We also thank Eastern Kentucky University and the University of Kentucky, for providing institutional support and release time on several occasions over the past years. Of course, all matters of data analysis and interpretation remain our sole responsibility.

PART ONE

mountain profiles

The Spiti River cuts through uplifted Tethys strata in the central thrust zone of the western Himalaya, near its juncture with the Sutlej River in northern India. Photo by D. Zurick

Himalaya under Siege

a SPIraL OF DarKISH COLOr appears on most world maps in the central part of Eurasia, depicting the greatest concentration of mountains on earth. Vivid, multihued arcs radiate outward from the Pamir range in present-day Kazakhstan, cross the boundaries of five countries, and make a tight knot of rugged terrain north of India. The bearing of these mountains on the maps makes a formidable cartographic statement, but that is nothing compared to their presence in the Asian landscape. The Tien Shan and Kun Lun mountains extend eastward from Kyrgyzstan and Tajikistan, through Turkestan and Tibet and across the Takla Makhan Desert, to wrap around the huge arid depression known as the Tarim Basin. The nineteenth-century travels of Swedish explorer Sven Hedin showed that area to be one of the most inhospitable spots on the planet. The Hindu Kush range, meanwhile, follows a southwestern trajectory through Nuristan, formerly known as Kafirstan—Land of the Infidels—to form the spiny backbone of austere and war-torn Afghanistan. The Karakoram peaks shape the icy ramparts of Pakistan, and tracing the southern border of the Tibetan plateau for 2,700 kilometers is the spectacular Himalayan range. The loftiest places in the world are located there.

Until a few million years ago, much of this region was flat and covered by the shallow water of the Tethys Sea. The layers of sediment deposited by that ocean, trapped now in the twisted strata of the Himalayan bedrock, show the successive advances and retreats of the primordial sea. The weathered mountain slopes expose an abundance of fossilized marine life—brachiopods, corals, and skeletal fish—that points to the locations of the old beaches, reefs, and mud banks. The fossils in the rock show that the ocean which once intervened between Tibet and India is ancient indeed; the outlines of early marine life imbedded in concretions of black slate are the remnants of organisms that lived during the Jurassic Age, when dinosaurs roamed elsewhere on the planet.

The tectonic collisions of the Indian and Asian plates, driven by the slow northward drift of southern Asia, raised huge sections of the old ocean floor, so the fossils are now trapped at the foot of glaciers and among river terraces situated many kilometers above the current level of the world's oceans. Seashells are found even on the summit of Mount Everest. Among the Tethys relics are the treasured ammonites, whose mandala shape is believed to represent the cosmic order of the world. They are still retrieved from their icy encasements by Hindu and Buddhist traders, who sell them to pilgrims as talismans. For such people, the rocks hold a power that comes from the curious mix of Himalayan geography and mythology.

The oldest rocks in the Himalaya actually predate the formation of the mountains. They derive from the great plains of southern Asia. The crystalline thrust sheets of the Himalayan peaks share a geological history with the billion-year-old Aravalli range, located much farther to the south in the Deccan Plateau of India. Pieces of the planet's original mountains, they were formed during the breakup of the 200-million-year-old supercontinent known to geologists as Pangaea. The roots of the Himalaya were thus in existence long before India started to drift northward from Gondwanaland in the far distant Cretaceous era, long before its collision with Tibet almost 40 million years ago.

The slow northward creep of the Indian plate caused massive buckling and folds in the ancient bedrock, diminishing the span of the Tethys Sea about 40 million years ago, until the ocean disappeared altogether. This movement north narrowed the divide between India and Asia until it became the staggering heights of newly formed mountains. The tumultuous upheaval of the region, first studied in 1851 by the Geological Survey of India, is all the more remarkable because it has never stopped. The current geophysical research shows that India continues to push northward against the inner reaches of Asia at a breakneck speed of about five centimeters a year, causing the Himalaya Mountains to continue to grow at the rate of a centimeter each year. In geological reckonings, that is earth moving very fast.[1]

The tectonic collisions of India and Tibet have shaped the world's premier mountain range. When the Karakoram peaks in northern Pakistan are included, the Himalaya contains all fourteen of Earth's summits higher than 8,000 meters. From the Indian plains looking northward, they appear as an impossible line of jagged white peaks set sharply against a steely blue sky. The mountains manage to look both solid and ethereal at once, almost like a *fata morgana*. Up close, the Himalaya is a rugged landscape, with towering rock walls, deep green valleys, knife-sharp edges, and flowing glaciers, and of an overall scale that is hard to reckon.

The Himalaya region

The relief range from high to low spots in the Himalaya is unsurpassed in the world. Where the Kali Gandaki River cuts between the Dhaulagiri and Annapurna ranges in Nepal, a vertical gain of seven kilometers is achieved over a linear distance of only twenty kilometers. The result is the deepest gorge on Earth. Other great cuts in the mountains are produced, west to east, by the flow of the Indus River in Pakistan, the Alaknanda River in the Garhwal region of northwestern India, the Karnali River in Nepal, the Tistha River in Sikkim, and the Yarlung Tsangpo River, which flows south from Tibet through the Namche Barwa Gorge to become the great Brahmaputra River. Those valleys have no parallel in the world. Their great depth accentuates the heights of the overtowering peaks (see frontispiece).

In the superlative terrain of world-class mountains, the Himalaya Mountains stand magnificently above them all. They appear to soar into the sky. Their majesty, though, is more than a matter of height. The Himalaya contains tremendous natural diversity, and a biological wealth that makes it one of the most treasured places on the planet. The snowy peaks, including famous Mount Everest, have no equal among Earth's other great mountain ranges. They command the attention of the entire world, for their haunting beauty and the challenges they pose to the human spirit. In the folds of the mountains live a diverse and tenacious people who are uniquely accustomed to the harsh conditions of the highland world. They have weathered the storms and adversities for centuries and have cut the mountain into neat, terraced farms and villages. The ancient cultures that reside in the Himalaya contribute immensely to the sense of timelessness which permeates the place.

But in recent years, alarms have sounded across the Himalayan range. Disturbing reports of environmental damage and the loss of traditional lifestyles appear with great regularity in the popular media and in the scientific publications. The accounts describe scenes of denuded and scarred hillsides, runaway floods, and impoverished villages. The rapturous visions of a faraway paradise—a Shangri La, so common among the early writings about the Himalaya—are displaced in the more recent reports by woeful tales of rich farmland washing away to the ocean, of an exploding human population, and of hillsmen fleeing from their impoverished homes to the industrial plains, where they seek jobs and new ways of living. The blame for all this misfortune is placed not on the mountains' tumultuous geology, which makes them especially prone to earthquakes, erosion, and floods, but on the villagers who inhabit them. It is thought by many people that the pressures of the native population on the land have caused the deterioration of the mountain landscapes. That is a controversial accusation. Our own geographical studies in the Himalaya, which encompass much of the time during which the major social transformations are thought to have occurred, show the mountains to be enduring places, the people to be resilient and coping, and the landscapes to be diverse. But that doesn't mean all is well in the mountains. Where significant damage is prominent, it issues from a perplexing mix of natural and human factors, not from any single cause. The overall impression is one of a tenuous balance. In some places, that balance has been upset.

Environmental Alarms

In his influential 1976 book, *Losing Ground,* Erik Eckholm of the World Resources Institute vividly warned of an impending ecological catastrophe in Nepal, which in his estimation would affect not only the millions of hill farmers living in that country but also the hundreds of millions of people who reside along the riverbanks in neighboring Bangladesh and India. His interpretation shows a desperate struggle to survive in the mountains, a downward spiral of ecologic and economic ruin, in which the mountain habitats are held hostage by the degrading activities of a rapidly growing human population.

> Popular image and reality rarely diverge so widely as they do in the case of Nepal. Most know the kingdom as the photogenic home of Mount Everest, as an exotic Shangri-la sprinkled with pagodas and quaint villages tucked away in the folds of the Himalayas. The facade of romance and beauty remains intact, but behind it are the makings of a great human tragedy. Population growth . . . is forcing farmers onto ever steeper slopes . . . Villagers must roam farther and farther from their homes to gather fodder and firewood . . . Ground holding trees are disappearing fast . . . Landslides destroy lives, homes and crops . . . Topsoil . . . is now Nepal's most precious export, but one for which it receives no compensation.[2]

Eckholm's devastating report, although restricted to Nepal, is not unique. Numerous media accounts in the 1970s and 1980s made similar forecasts of impending environmental collapse all across the range. Eleven years after *Losing Ground* was published, *Newsweek* magazine, in 1987, reported on degraded conditions across the Himalaya: "In Pakistan, India, Nepal and Tibet, deforestation has eroded fertile top-soil from the hills, triggering landslides and clogging rivers and reservoirs with so much silt that they overflow when they reach the plains of the Ganges . . . At the rate trees are being felled for fuel and cropland, the Himalayas will be bald in 25 years."[3]

The early predictions have not yet come to pass; the Himalaya are not yet bald of trees. But the media reports continue to describe an acute environmental crisis in the mountains. The focus of such accounts is on the degraded status of the alpine forests and on the contribution of deforestation to such problems as soil erosion and flooding in the lowland plains. Indeed, forest cutting has become synonymous with land degradation across the Himalaya. The environmentalists in India, for example, use the slogan, "Dam the Himalayas with forest cover, not with engineering dams," to voice their direct opposition to large hydropower dams and to timber cutting in the Kumaun and Garhwal mountains of the western Himalaya.[4] They elicit concern over the loss of productive farmland, the abnegation of village control over natural resources, and the overall fragility of the mountain ecosystem.

Such portrayals render a powerful image of the Himalaya under seige, compelling not only the actions of ardent environmentalists throughout the world but also the contributions of hundreds of millions of dollars in foreign aid and the rush of legions of foreign tourists to visit the mountains before they indeed wash away. The ecocrisis scenario has proven highly successful at generating revenue for the regional economy. The Himalayan countries, in turn, have become highly dependent upon the largess of the international donors. But do the portrayals of imminent environmental collapse best reflect the current conditions in the mountains, or do they impose a green myth, largely unsubstantiated, as many people are inclined to think?

It is clear that widespread disturbances in the alpine landscapes are indeed taking place, but it is not yet certain whether the changes necessarily lead to degraded places. The Himalaya Mountains, after all, have been settled for centuries by people who shaped the land to meet their needs. Terraced farms were cut into the hillsides hundreds of years ago, forests were cleared for homes and livestock pasture, streams were diverted for irrigation and to power small grain mills, rocks were quarried for walls and temples. So the lay of the land has taken on an increasingly anthropomorphic dimension. Until fairly recently, the question of land degradation did not exist at all in the Himalaya.

But the number of people living in the mountains has trebled since the 1950s, causing ever-greater demands for food, fodder, and fuel wood. The industrial designs on the Himalayan resources have also accelerated, especially for timber and water, so the range of human pressures on the land is both wide and deep. The rate and intensity of the recent landscape changes far surpass those that occurred in the past, prompting widespread concern about the sustainability of the mountains as natural areas and as places for human life. The clearing of the Himalayan forests is also thought to cause severe flooding in the densely settled plains of India and Bangladesh. That is the so-called highland-lowland dilemma of the Himalaya. It forges a bond between the life in the uplands and that in the valley of the Ganges River. The actual mechanisms are complex but, simply put, the degraded forests may hold less soil, catch less runoff, and stabilize fewer slopes. That all appears to cause the higher sediment loads in the rivers, which then surge in unpredictable floods onto the Ganges plain, sweeping away villages, livestock, and precious farm soil.

In this alarmist view, the linkage between the Himalaya Mountains and the southern Asian lowlands is a detrimental one. The mountain farmers are seen as the culprits who upset the ecological balance all the way from the plateau of Tibet to the Bay of Bengal. Most of the popular reports, though, document only short-term visits to a few selected mountain localities. They tend to focus on the subsistence needs of the farmers, rather than on the extractive industries of the commercial economy. And they assume, too simplistically, that the visible changes to the landscape automatically lead to land degradation. But those observations are commonly extrapolated across the mountain range to describe conditions throughout the Himalaya. This high level of generalization is misleading; it discounts the tremendous environmental and cultural diversity that characterizes the mountains.

Other accounts, though, rely on more detailed scientific measurements and on long-term formal studies to describe changes in the alpine habitats; those are not so easily dismissed. Many scientific studies also show serious forest loss or degradation, depleted farm soils, overgrazed pastures, and alarming levels of human poverty. They suggest worsening conditions for life in the mountains. But instead of seeking a simplistic equation between human population growth and land degradation, as in many of the popular media accounts, and instead of promoting the questionable linkage between alpine deforestation and lowland floods, the scientific studies relate the overall problem of land degradation to the deeply intertwined problems of subsistence pressures, land tenure, commercial resource extraction, industrial development, roads, and tourism. In our estimation, their potential contribution to environmental damage in the Himalaya is most important.

The various popular and scientific reports are so contradictory that the status of the Himalayan environment remains uncertain and contentious. Indeed, some writers have even penned the environmental crisis in the Himalaya to be a matter of words, in which emotion rather than fact underlies our notions about Himalayan deforestation and ecological change.[5] In the influential 1989 book *The Himalayan Dilemma*, the diverse Himalayan reports are synthesized into a grand theory about land degradation. The authors, Jack Ives and Bruno Messerli, argued that the Himalayan ecocrisis is untenable, a myth based upon popular but unsupported assumptions, suspect data, and gross generalizations. It is a serious indictment of the prevailing Himalayan environmental model and a concerted effort to set the record straight. According to Ives and Messerli,

> It is not our intention to dispute the facts, wherever reliable information exists, but the assumptions that so frequently are not based upon facts. Nevertheless, throughout this attempt to dissect the Theory of Himalayan Environmental Degradation the causal relationships between timing and degree of population growth, deforestation, loss of agricultural land, and downstream effects are paramount. We will attempt to demonstrate that most of these linkages and assumptions are founded upon latter-day myth, or falsely based intuition, or are not supported by rigorous, replicable, and reliable data. They are the 'sacred cows' of the perceived

Himalayan Problem, and we will seek to dismantle them, in part or in whole.[6]

The Himalayan Dilemma readily shows how the portrayal of a Himalayan "supercrisis" fails to accurately capture the unique conditions of nature and society in the Himalaya. It exposes the untenable linkage between deforestation in the mountains and the incidences of heightened flood damage in the plains of southern Asia. And it highlights the fact that overgeneralization is always counterproductive to understanding the nature of environmental problems in such a diverse place as the Himalaya Mountains. But despite the book's worldwide influence, as well as the efforts of many follow-up studies, speculative and conflicting images of the region persist. Scholars, politicians, and visitors continue to see the mountains in significantly different ways, and differently from the views of native residents. Furthermore, because the perceived status of the environmental problems remains contested, there is no holistic design of reasonable policies to resolve them.

The answer to this puzzlement lies partly in the point of view of the chroniclers. If persons seek a single vision of the Himalaya, they will remain frustrated, for there is no such thing. The cold arid highlands of Zanzkar and Spiti, located in the far northwestern region, are worlds away from the warm orchid-filled forests of Sikkim or the thatched homes of the Tharu tribes, who live in southern Nepal. None of those places much resemble Nepal's bustling capital city of Kathmandu either, or the other industrial towns that have sprung up across the lower mountains. The Himalayan range is one of the most disparate places on Earth, and no single description will capture the full range of its qualities. Hence, the general scenarios of environmental collapse that so commonly appear in the literature are flawed; the reality they may catch is as fleeting as the mountain clouds.

As a result, if the policy makers look for a single model of the Himalaya to facilitate their development programs, they too will not succeed. For just as the mountain ecosystems and cultures shift rapidly across the rugged terrain, so too must new social and economic programs accommodate the local exigencies of geography and history. On the one hand, this denies the assumptions of an essentially Himalayan problem, as portrayed in the numerous accounts of imminent ecolog-

Landscape of cultivated fields and barren mountainsides in the Kali Gandaki Valley, between Annapurna and Dhaulagiri ranges. Photo by D. Zurick

ical collapse. But the high degree of diversity found in the Himalaya does not vitiate the existence of serious environmental and social problems in the region; indeed, these problems are more complicated than earlier imagined, and perhaps of a greater magnitude. If a regional scenario of land degradation were to hold true, then a general prescription for its resolution might well succeed. Neither, as it turns out, is the case.

This, then, raises the question of how the Himalaya might be understood as a region, which is precisely what this book attempts to answer. The heterogeneity of the mountains may be, after all, their most compelling characteristic; but despite its internal diversity, or rather *because of that diversity*, the Himalayan region contains, in our view, a complementary geography. Much like a creative mosaic of independent but integral components, the Himalaya is a contiguous product of nature and society. The events that shape the region cut across the grain of the mountains, giving them a cohesion that transcends the look of the land.

Himalayan Motifs

The settled places in the Himalaya reflect various human needs, aspirations, and imaginings. Land and life coexist there in multifarious, but often troubled, ways. On the one hand, the traditional alpine societies look inward and to their pasts in order to locate the best models for their continued survival. That is their conservative nature. But the villagers must also accommodate the political and economic forces that emanate from societies that lie out of reach, beyond the intimate spaces of their lives. Such accommodation has, to varying degrees, always been the case, but the pace of change has quickened in recent decades. Modern national and regional developments see the villages as promising settings for the commercial conditions of life. Rural industries are widely promoted to augment the local economies, increase the levels of national productivity, and alleviate some of the region's material poverty. These are laudable goals. But the industrial expansions supplant much of the traditional livelihood and shape entirely new mountain landscapes. They herald a new order for the places that lie at the very edge of the world.

The various agricultural schemes, for example, such as the new crop developments and the establishment of orchards, show a steady march toward the marketplace. They introduce to the agrarian economy the high-yielding grains, machine-based processing methods, and intensive land management that characterize industrial farming. The new factors of production literally transform the mountains as they expand the size of farms, redesign irrigation schemes, and introduce fossil fuels and chemicals to the previously organic farming systems. The intention of such innovations is in part to promote food production in order to keep pace with the rapid population growth. But they also seek to increase the overall economic output of the region for purposes of national development.

The horticultural programs, now widespread across the temperate zones in India and Nepal, constitute an important type of commercial development in the mountains. Apple orchards, with their attendant need for access roads, nurseries, and processing sheds, already cover 325,000 hectares in the northwestern Indian Himalaya alone. The orchards are a highly specialized land use in localities where growing conditions are especially suitable for fruits, or where historical events contributed to a local fruit industry. They now claim 20 percent of the total cultivable land in the mountains of Himachal Pradesh.

Apples were first introduced to the Himalaya in the 1930s, by an American missionary living in the small village of Kotgarh, above the Sutlej River near the town of Shimla. At that time, Shimla was a prosperous British hill resort. The American apparently meant to surround himself with a familiar and productive landscape, but apples caught on in a big way among the local people. With government support, the orchards have since spread all across the slopes of the western mountains, and are especially notable along the Sutlej and Beas rivers and their numerous tributaries. They give that part of the Himalaya a different look and new prospects for economic growth.

Many of the developments taking place in the remote areas of the Himalaya reflect the priorities of national development rather than needs of local villages. They force the villagers to accept new sets of social values and economic aspirations. That is a logical outcome of modernity, but it often comes at considerable cost to the mountain communities. For example, the new programs designed to exploit forests, farmland, water, and other natural resources for commercial gain may ignore the natural capacities of the environment or the subsistence goals of the villagers. They require new appraisals of mountain resources, ones that may be at odds with traditional values and sustainable land uses. The security of land tenure, which is vital to the well-being of native residents all across the Himalaya, may disappear altogether in places where the local rights over the land are usurped by the new forms of environmental management. Such conflicts are already widespread in the region, for example, where parklands are designated without the advice of villagers, or where government industries have become established and profoundly change the economic patterns of community life, intimating new ecological futures for the mountain landscapes and people.

The Himalayan peoples nonetheless hold on to their own unique cultural identities, forged from long-standing appraisals of the conditions of mountain life. The enduring relations between nature and society may be threatened by the rapid intrusions of economic and social programs, but they are not destroyed by them. Numerous examples show how villagers will fight to keep what has been theirs for centuries. The indigenous land struggles, most often waged by villagers against the advancements of corporate interests, sometimes find the support of governments. In Bhutan, for example, the native efforts to maintain cultural authority over land are encouraged and supported by formal national policies. The development plans of Bhutan are written explicitly to perpetuate the national culture and to promote the environmental goals that are central to the country's Buddhist ethics. In so doing, they benefit from centuries of dharmic tradition.

The resiliency of mountain cultures can be seen also among the Sherpa people living amid the spectacular peaks of the Khumbu region in eastern Nepal. The Sherpa have managed to keep intact much of their unique tribal identity, despite the worldwide attention on nearby Mount Everest. Farther to the west, in Nepal, the Annapurna Conservation Area Project, known worldwide by its acronym ACAP, administers 7,000 square kilometers of mountain land in conjunction with the local villagers. The goals of ACAP are to preserve the environment and to simultaneously promote economic development in the villages. Futhermore, both objectives are linked to maintaining the high degree of cultural diversity in that region. The commitment of the Annapurna Project is to keep alive traditions while enabling people to prosper amid the changing political and economic landscapes: a difficult undertaking.

But people have resided in the Himalaya for a very long time; they know the mountains well, have put their mark on them, and have staked a vital claim to them. The land is the basis for their livelihoods. Mountain communities living all across the range continue to resist the development pressures that might otherwise jeopardize their access to land and alienate them from their mountain homes. Such struggles are readily seen in the efforts of the tree-hugging women of the Chipko Protest, which has become world-famous as a model of grass-roots activism, and of the Tehri Dam opponents in the mountains of Garhwal in India. Those are well-documented activist movements and have international support. It is found also among numerous other, less-publicized, indigenous land struggles located elsewhere in the Himalaya. The environmental battles endow the mountains with the strength of their native cultures and counter the potentially divisive forces of irresponsible development by maintaining villagers' authority over resource decisions. The mountain communities thereby retain a sense of dignity amid the sweeping transformations caused by modernity. Environmental conflicts may be resolved; cultures may flourish. Under such circumstances, the Himalayan landscape is reflexive and ever-changing, but it remains mainly under the control of those who live there permanently. Elsewhere, though, where national or corporate interests have come to dominate and alienate local people, loss of culture and degradation of the environment are commonplace.

The much-publicized "opening up" of the Himalayan borderlands to foreigners—as occurred in the early 1980s in Ladakh, in 1992 in Dolpo and Mustang in Nepal, and in 1994 in Sikkim and Spiti—has fostered the notion that such isolated places were formerly without much outside contact. That is not the case. The mountains do not create such an impenetrable barrier for their residents, nor are the people so disinclined to travel. The history of human settlement in the Himalaya is marked, instead, by regular mobility for purposes of trade, resources, work, pilgrimage, or social exchanges.[7] The intricate networks of walking trails, resting places, and mountain passes, and the cultural traditions of innkeeping and porters, confirm that travel has been an ongoing feature of the mountain worlds. Trade, livestock movements, journeys to work and ceremony are all traditional events that link the Himalayan places to one another and to the outside world.

From such a viewpoint, we can begin to imagine how the Himalayan communities may greet the new demands of wider society.[8] It is not a smooth transition. The traditional forms of resource management are authorized by systems of power, prestige, and entitlement, though they are not exclusively political. Ritual and belief, spontaneity, ethics, and experience, all continue to anchor people in their place and serve as guides to life. But new economic orientations occur when the villages join with the wider region, predicting new relations of power and wealth, and different ways of evaluating resources. The clash between tradition and modernity, a defining feature of remote places the world over, sparks a new phase in the Himalaya, when alarming disjunctions may occur between people and the places where they live.

Concrete examples of this unsettling phenomenon include the breakdown of traditional systems of knowledge and the loss of land to outside capital investments; the growing poverty of hill farms and the despair that it gives rise to; and, perhaps most tellingly, the stream of people who leave their Himalayan homes to seek new lives in entirely different places. In Nepal alone, for example, over one million people are on the move each year. Most of them leave the mountains for the city lights of Kathmandu, the promise of open land in the tarai (piedmont plain), or the possibility of seasonal work in India. The widespread migration of people from the highlands to the lowlands is a compelling sign that something is seriously amiss in many of the mountain villages.

But such downward trends are not the only or necessary case in the region. In recent years, and largely at the urging of the mountain people themselves, more sustainable living designs are proposed which hold a greater promise for the futures of Himalayan communities. For the most part, they are locally based programs, meant to provide new economic opportunities in the rural areas by maintaining a viable relationship between the villages and their surrounding environment. The new programs help to keep mountain people literally in their place. As Derek Denniston, of the Worldwatch Institute, wrote about the Himalaya, "Across the range, locally-directed projects are restoring the self-sustaining nature of the mountain economy and ecology. By placing natural resources back under the control of villagers, these projects give local people both the incentive and the means necessary to break the interlocking grip of poverty and environmental decline."[9]

Although many such programs are only now being imagined, several ongoing projects already show the possibilities. The new Himalayan parklands in Nepal and Bhutan, for example, blend native ways of managing resources with new institutional guidelines to conserve forests and to protect critical habitat for endangered plant and animal species. The parks attempt to match biodiversity management with economic programs, such as nature tourism and handicrafts, which may then provide opportunities for local people to earn incomes while at the same time acting to preserve the mountain environments. Such efforts are not without their problems, but where they are successful the innovative parks show the value of involving local residents in their design and management.

The park concept has taken hold in a big way all across the Himalaya. In northwestern India, more than sixty parks and sanctuaries have been commissioned. Twelve major national parks and reserves are located in Nepal. Bhutan's *Seventh Five-Year Plan (1992–1997)* sponsors twelve protected areas covering 20 percent of that country. There are now 130 protected areas in the Himalaya region, covering some 13,600 square kilometers. The largest Indian biosphere reserves, located at Nanda Devi and in the Valley of the Flowers in Uttarkhand, contribute 6,500 square kilometers of protected land to the Himalayan list. Much work remains, however, to insure that the protected areas are more than merely paper parks on the planners' hopeful maps. Their full implementation requires considerable financial investment, as well as diplomatic efforts to gain the goodwill of local residents.

Meanwhile, people living outside the mountain parklands are also devising new ways of managing local development to keep within cultural traditions and natural carrying capacities. Many such programs put environmental resources directly under the control of rural communities.[10] Oftentimes, the local programs result from long-term struggles by villagers against the interventions of outside operators. The Chipko Protest, for example, began in the 1970s in the mountains of Garhwal in India, when a group of women linked arms and encircled the village trees to keep them from being cut by the commercial lumbermen. A sports equipment company wanted to turn their forests into tennis rackets; the villagers had other plans. The ultimate success of the movement gave Chipko a worldwide reputation as an activist struggle, but it gradually developed into a more comprehensive guide for sustainable development throughout the mountains.[11]

In other cases, the local alternatives to high-impact development schemes have gained the support of government agencies. In limited instances, the big developments are recognized by politicians to be unsustainable for the country. Some Himalayan countries have installed entirely new agencies to manage grassroots-based environmental conservation. In Nepal, the King Mahendra Trust for Nature Conservation designs and manages local development projects in the country and encourages villagers to participate in the design of future programs. Bhutan has taken a kindred approach with its Bhutan Trust Fund for Environmental Conservation. The trust seeks to manage the financial commitments necessary to preserve the kingdom's environmental future. In Sikkim, the local efforts to decentralize planning have helped to foster greater participation in development by the marginalized bhotiya and Lepcha peoples. Finally, in the far northeastern Himalaya, among the hill tribes of Arunachal Pradesh, several participatory development programs are under way. Those efforts, although still quite limited in their scope, recognize that local people must be engaged in the design of their future worlds if sound development is to occur in the Himalaya.

The Brahmaputra River, which forms the eastern divide of the Himalaya, is more than a mile wide at the town of Pasighat in Arunachal Pradesh, India. Photo by P. P. Karan

Mountain Contours

THERE IS NO REALLY COMPELLING geological argument to show clearly where to divide the Himalaya from adjoining mountain ranges. They are all part of a huge knot of mountains that is hard to fully disentangle. Consequently, the placement of boundaries to shape the region is a problem of geographical interpretation. Some measurements of the Himalaya include Afghanistan's Koh-i-Baba range in the west and the highlands north of Burma in the east, making the Himalaya over 4,000 kilometers long. That would extend the mountains across eight countries, through the Hindu Kush and Karakoram ranges to the Hengduan Mountains in the Yunnan province of China. More conservative maps put the western limits of the Himalaya at the Oxus River and omit the Hindu Kush, thus restricting the range to about 3,000 kilometers. We distinguish the Himalaya from the

Karakoram range in the west and the Burmese highlands in the east by using the bends of the Indus and Brahmaputra rivers, respectively, to mark the western and eastern boundaries. In that estimation, the Himalaya Mountains cover a distance of twenty-two degrees of longitude, or some 2,700 kilometers on the ground, and contain areas in Pakistan, India, Nepal, Tibet, and Bhutan.

The northern and southern boundaries of the Himalaya likewise are not firmly fixed. The Indus and Yarlung Tsangpo rivers contain the northern slope of the mountains as they flow west and east from their headwaters beneath Mount Kailas in Tibet. Most of the Tibetan plateau is separated from the Himalaya by those magnificent rivers. The plateau country is structurally a different landscape and it is easily distinguished by the human eye when viewed against the sharp rise of the high peaks. At close range, the dividing line between the mountains and the Tibetan steppeland is a chaotic landscape of deep gorges, steep valleys, and high passes situated between towering summits, and not a straight edge at all.

The great peaks of the Himalaya pose a formidable natural wall south of the plateau. These high mountains descend southward in a series of lesser ranges to eventually reach the lowland plains of the Ganges River, where the mountains meet the Indian tableland in a foothills zone marked by frequent seismic tremors and earthquakes—one of the most active tectonic regions in the entire Himalayan range. The southernmost extensions of the mountains compose a piedmont plain, a broad, gently sloping landscape interrupted occasionally by hills and broad river valleys.

In this general overview, the Himalaya contains an unparalleled landscape. At its westernmost extension, the range is anchored by the rock-faced summit of 8,125-meter Nanga Parbat. The Pakistan districts of Mansehra, Abbottabad, and Kohestan all lie within the shadow of that great mountain, held there by the turbulent reach of the Indus as it bends south and then west at the base of Nanga Parbat. Arcing toward the southeast, the Himalayan range forms the Indian territories of Kashmir, Garhwal, and Kumaun. The territory covered by the mountains in northwest India is 350,000 square kilometers, or more than one-half of the total Himalayan region.

The loosely guarded border between India and Nepal follows north along the Maha Kali River, joining the Garhwal territory of India with Nepal's western frontier in a region that shares similar cultural traits and environmental conditions. The Himalaya continues eastward from the Maha Kali across the entire breadth of the kingdom of Nepal, containing in that country a total area of 147,000 square kilometers. The range again enters India in Sikkim and Darjeeling, where it composes 10,450 square kilometers of land wedged between Nepal and Bhutan and dominated by Mount Kanchenzonga, the third-highest summit in the world and one of the range's most beautiful mountains. All of Bhutan's 46,500-square-kilometer area is included in the Himalaya. The remainder of the range, about 52,500 square kilometers, straddles the Assam Mountains of Arunachal Pradesh, in India's remote Northeast Frontier.

The far eastern edge of the Himalaya is formed by the southward bend of the Brahmaputra River. Known in Tibet as the Yarlung Tsangpo as it crosses the plateau, the river shifts south at about the 95° East Longitude line, looping around the 7,756-meter Namche Barwha and cutting a great gorge through the mountains. It formally changes its name to Brahmaputra near the town of Dibrugarh, north of the Assam Valley.

In its widest stretch, from the Indus catchment in the west to the Brahmaputra valley in the east, and from the outer margins of the Tibetan plateau in the north to the lowest foothills and adjoining plains

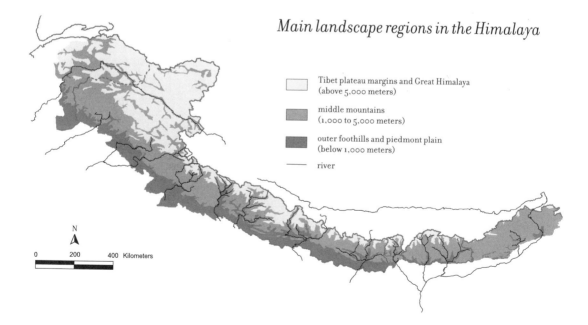

Main landscape regions in the Himalaya

Tibet plateau margins and Great Himalaya
(above 5,000 meters)

middle mountains
(1,000 to 5,000 meters)

outer foothills and piedmont plain
(below 1,000 meters)

—— river

N

0   200   400 Kilometers

in the south, the Himalaya measures about 625,000 square kilometers in area. The scale of the mountains is immense; the summits are magisterial. Overall, the Himalaya composes an exquisite mosaic of mountain places, each unique in terms of geography and human society. The regional divisions of the Himalaya take account of how physical factors such as topography, climate, and soils control the distribution of natural vegetation and environmental productivity.[1] Some classifications of the mountains are based also upon cultural patterns, recognizing the important fact that 50 million people live in the Himalayan region. Their presence overwhelms the simple geographical models by adding the diverse dimensions of human settlement, history, and economy.

A few fully nomadic communities live in the dry plateau zone, in places such as Ladakh and Zanzkar. Their numbers are low and further declining as pasture areas are lost to other economic purposes. In the moist high valleys of the Great Himalayan zone—for example in Langtang and Khumbu in Nepal or in Kashmir—the people move seasonally with their livestock, looking for grass and water. They travel to the high pastures in the summer months and return in the winter to their permanent farming villages located at lower elevations. The main agricultural areas are distributed widely across the lower, rainfed southern slopes of the middle mountains, all the way from Kashmir to Arunachal Pradesh. The rugged terrain, abundant microclimates, and diverse cultures found in the middle hills predict a galaxy of natural conditions and cultural adaptations. Finally, the outer piedmont plain supports fertile agricultural villages, new industrial towns, and wildlife refuges.

Sketches of the Himalayan Landscape

The Himalayan environment, compelling in its topography, is also exceptionally rich in biological life. The great diversity of fauna reflects the fact that the mountains straddle the major Paleo-Arctic and Indo-Malayan life zones, hosting species that originate in both those major earth realms. An abundance of ecological niches occur which support unique and complex evolutionary pathways for native flora as well. For example, over 650 different kinds of orchid thrive in the wet mountains of tiny Sikkim. They share the forests with numerous local species of tree ferns, bamboo, and rhododendron and with the red panda, brown bear, and silver foxes that roam the forests. The rapid elevation changes in Sikkim allow plants from numerous climate zones to grow profusely together in a riotous mix of habitats replicated nowhere else on earth.

At a high level of generalization, the climate types and vegetation that one encounters in the vertical relief of the Himalaya Mountains is analogous to the range of environments one might expect to find by traveling from the subtropics to the polar ice fields. It is quite possible to journey in a single day on foot between the tropical and temperate worlds. The geographic isolation of the mountains also gives them a special role as sanctuaries for many endangered or threatened plants and animals. High rates of species diversity and endemism—that is, where organisms are found nowhere else—make the Himalaya especially susceptible to the loss of entire natural communities and hence to imperiled biodiversity. Nonetheless, measured against the great environmental transformations that have occurred in the adjoining lowlands, the mountains stand as a unique planetary biome.

The preeminent characteristic of all mountains is their verticality. The rapid changes in altitude encountered among alpine settings correspond to significant differences in local climate. Because the relief ranges are especially great in the Himalaya, where it is not uncommon to traverse tens of thousands of feet of altitude change in only a few miles of ground distance, the vertical dimension is particularly outstanding. It combines with other geographical factors such as topography, orientation to the sun, and the presence of glaciers and meltwater to create the complex distribution of Himalayan life zones.

The interplay of physical factors—soils, water, climate, and terrain—is followed closely by the natural distribution of plant communities. Hence, mapping vegetation patterns is a useful way of distinguishing environmental zones in the Himalaya.[2] Vegetation maps tend to be oriented along the parallels of latitude, such that the forest types appear to climb the mountains as if on a ladder. Because human endeavors capitalize on the verticality of mountain worlds, the cultural use of land must also be considered. Settlements, trade routes, farming regions, and livestock pastures add even greater complexity to the contours of the Himalayan worlds.[3]

Apart from the role of altitude, latitudinal position and the occurrence of the Asian summer monsoon are reflected in the mountain life zones. In order to fully capture the range of landscapes found in the Himalaya Mountains, it is necessary to consider all these factors. The Indian regions of Ladakh, Kashmir, and Garhwal occupy the northernmost position in the Himalaya and therefore exhibit the lowest temperatures for similar elevations anywhere in the mountains. Assam, meanwhile, enjoys a warm climate all year round, owing to its southern location.

Whereas geography accounts for most of the temperature differences found in the Himalaya, precipitation has a decidedly temporal nature. A gradual warming of the interior of Asia occurs with the early summer months, creating a huge low-pressure system above northern Tibet that draws moist air northward from the Bay of Bengal and the Indian Ocean. The airflow cools as it rises over the Himalayan peaks, condensing the water vapor so rain or snow falls in the mountains. The southern slopes of the range receive the bulk of that precipitation, while the lands that lie to the north of the mountains, in the lee of the moisture-bearing monsoon winds, are in a rainshadow and remain a virtual desert. The overall precipitation also drops from east to west as the influence of the monsoon tapers off. The greatest rainfall receipts are recorded on the southern aspects in the eastern portion of the range, with some spots averaging more than ten meters of rainfall a year. One of those places is Cherapunj in Assam, which may be the wettest spot on the planet. Cherapunj gets more than fifteen meters of rain a year, most of it falling during three summer months.

The broad climatic belts that relate to altitude, latitude, and the monsoon exhibit a great deal of local variability. For example, the cyclonic storms that sweep down from central Asia into the northwestern Himalaya during the winter months bring much-needed winter rain to Garhwal, Kumaun, and western Nepal. The eastern range, meanwhile, stays dry during the winter. The local topography also influences the amount of rainfall, sunlight exposure, and the strength of winds. Pockets of drought, produced by intervening ridges, create the classic "dry valleys" in the Great Himalaya. The south-facing slopes of the mountain valleys tend to receive on average several hours more sun each day than the shadowed northern slopes. In consequence, while the distance between points on two facing slopes may be only a few hundred meters, the change in vegetation or in agricultural potential may be quite great.

Notwithstanding these important local variations, several reasonable divisions of the Himalayan landscape may be drawn according to climate and terrain: the southern margins of the Tibetan plateau, the Great Himalaya, the middle mountains, and the outer foothills and piedmont plain (in Nepal known as the tarai and in Bhutan as the Duars Plain). These divisions strike along an approximate northwest to southeast axis and extend from the Tibetan highlands in the north to the Indian plains in the south.

THE SOUTHERN MARGINS OF THE TIBET PLATEAU

With the notable exception of Ladakh in the far western sector of the Himalaya, the Tibetan plateau covers a relatively small portion of the mountains proper. The plateau country mainly occurs along the extreme northern edge of the region, where the geological formations and the rainshadow effect of the Great Himalaya produce a steppe landscape similar to that found across much of Tibet. Ladakh is the largest of such areas, containing about 60,000 square kilometers of arid land along the upper Indus River. To the south of Ladakh, across the eastern portion of the Zanzkar range, lies the valley of the Spiti River, an important tributary of the Sutlej River system. The Spiti valley shares a common border with Tibet, contains about 6,000 square kilometers, and averages over 4,500 meters in elevation. It is a windy, dry, and starkly beautiful landscape that contains only a few areas suitable for agriculture and livestock pasture. Most of the region is without water and is empty of permanent human settlements.

Farther to the east, in the Nepal Himalaya, are the plateau borderlands of Dolpo and Mustang. The latter, known historically as the kingdom of Lo, is a southward extension of Tibet's Tsangpo Plain. It contains the headwaters of the Kali Gandaki River and functioned as an independent principality as recently as ten years ago. Dolpo is situated to the west of Mustang, in the rainshadow of the Kanjiroba (6,882 meters) and Dhaulagiri (8,167 meters) massifs. Additional southern extensions of the Tibetan plateau occur in Bhutan north of Kula Kangri (7,320 meters), in Sikkim at the upper Tistha River catchment, and along a narrow strip in the Assam Himalaya, where the Indian state of Arunachal Pradesh borders Tibet.

Throughout the plateau zone, the snowy peaks overshadow a broadly folded terrain of sedimentary mountains and intervening river valleys. The altitude of the zone varies quite a bit, but averages between 4,000 and 5,000 meters above sea level, with many higher peaks. The plateau's snowfields and glaciers form the principal headwaters of the main Himalayan watersheds. Hence, this remote region is critical for the delivery of rivers to vast areas located south of the main Himalayan crest. Situated as they are in the lee zone of the main mountains, and hence beyond the reach of the summer monsoon, the plateau landscapes themselves remain dry throughout the year. This zone supports only limited natural vegetation. Grasses, thornbushes, and scrub junipers are most common. Farming occurs where irrigation is possible, limited to fertile oases along the major streambeds and river terraces. Buckwheat, barley, and potatoes are important crops grown in those places. Yak, goats, and sheep are kept by the villagers living in valley oases, and encampments of nomadic herders are found among the grassy steppes of the region. The farmers and herders trade grain and livestock in centuries-old systems of barter.

The remote terrain of the plateau zone makes communication and travel across the Himalaya difficult, but several high passes lead into the zone and provide important trading routes. The major routes include the Jelep La (4,373 meters; *la* means "pass"), from Sikkim to Lhasa through the Chumbi Valley; the short but high route across the Laitsawa Pass (5,150 meters), in Bhutan; the high routes between Nepal and Tibet located in Khumbu, Mustang, Dolpo, and Humla; and the crossings of the Zanzkar and Ladakh ranges at Baracha La (4,650 meters) and Kun Zum La (4,276 meters). These passes link isolated plateau settlements with communities in other valleys and allow trade caravans to move south and north across the Great Himalaya. A few of the passes are now crossed by motorable roads as well, so buses and trucks travel the high routes for a few months each year when the roads are free of snow.

THE GREAT HIMALAYA

South of the plateau zone are the highest peaks of the Himalaya, where Tethyan deposits over fifteen kilometers thick have been uplifted to expose the crystalline core of the mountains. The Great Himalaya span the entire length of the range, from Nanga Parbat in the

The southern margins of the Tibetan plateau lie along the India-Tibet boundary. It is an expansive arid region of deep valleys and magnificent folded rock cliffs. Photo by D. Zurick

The Great Himalaya peaks located in the snowy Zanzkar range in India tower above the surrounding plateau country. Photo by D. Zurick

west to Namche Barwa in the east. The region is dominated by a vast stretch of snowy summits, glaciers, and rock walls, interrupted only occasionally by deep gorges or high saddle passes. Secluded tributary valleys intervene among the summits and provide protected places for human settlement. The renowned summits of the Himalaya are all located here: the pyramidical Nanda Devi (7,816 meters) in India; the giant massifs of Annapurna (8,091 meters), Manaslu (8,156 meters), and Everest (8,848 meters) in Nepal; and the exquisite Mount Kanchenzonga (8,598 meters), which straddles the border between Sikkim and Nepal.

The Great Himalaya zone is sparsely populated. The extreme cold, snow, and steep slopes prohibit most forms of human activity. The villages are restricted to the larger river valleys and to the south-facing slopes where soil temperatures are adequate to support agriculture. In Zanzkar, for example, which encompasses the northwestern portion of the range, people live only along the upper stretches of the Zanzkar, Stod, and Lung-nak rivers. Their cloistered settlements are located near the Buddhist monasteries and close to sources of water. The clear skies that prevail in Zanzkar throughout much of the year provide sufficient sunshine and warmth for farming in the summer months. Cultivable land extends along the alluvial river terraces, where irrigation systems have been built to bring water to the fields. Satellite settlements, called *doksa*, are located in the higher grazing areas, but are occupied only from June to September by nomadic herdsmen.

The winters in Zanzkar are long and the snowfall is heavy. People who have not left the region for work or extended family visits are restricted to their homes, where they engage in handicraft work, weave, and complete the tedious rounds of daily chores. A variety of religious and social activities punctuate the highlanders' lives, though, and help to regulate the ways in which villagers use the meager natural resources of this harsh environment. Each year in the dead of winter, when the Zanzkar River is iced over, a band of traders follows the frozen river to the roadhead leading to Ladakh. It is a treacherous journey, taking over a week, but it is the only way for the villagers to leave the valley in the winter.

In recent years, some places in the Great Himalaya zone have attracted a clientele of international tourists who come to trek on the mountain paths and to view the rugged alpine scenery and the ancient cultures. The strong appeal of this zone for tourists predicts significant changes in the future ecology and economy of the Great Himalaya. Where the tourists do not go, though, the patterns of life in the remote high valleys unfold in some of the most timeless ways found in the entire Himalayan range.

THE MIDDLE MOUNTAINS

The numerous ridges that radiate southward from the Great Himalaya crest form the backbone of the temperate middle-mountains zone. It is the largest and most densely settled part of the Himalayan range. The middle mountains are actually quite high when compared to other world mountains, averaging 2,000 to 4,000 meters in elevation, with some ridgetops over 5,000 meters. The terrain is heavily dissected by rivers, so the region overall exhibits a rugged topography with many discrete mountain ranges. The Pir Panjal in the northwestern Himalaya, the Mahabharat Lekh in Nepal, and the Black Mountains of Bhutan are prominent examples of high ridges located within the middle-mountains zone.

The valleys that cut through the middle mountains are formed by tributaries that originate among the glaciers and snowfields of the Tibetan plateau and the Great Himalaya. They converge to form several large river systems. From west to east, these are the Indus-Sutlej and the Alaknanda-Bhagirathi systems in India; the Karnali, Narayani, Gandaki, and Kosi rivers in Nepal; the Tistha in Sikkim; the Amo-Sankosh and Manas river systems in Bhutan; and the Brahmaputra River in the far eastern sector. Their extensive drainage systems play a crucial role in shaping the topography of the middle mountains. They cut the land into a striking mosaic of knife-sharp ridges and deep valleys and gorges. The rivers also provide important irrigation water for farming in the zone as well as the potential for generating hydroelectricity.

Because it is situated south of the barrier formed by the high peaks, the middle-mountains zone receives a great deal of precipitation during the summer monsoon months. The annual rains give much of the zone a moist, temperate outlook. Where human activities are restricted, such as in parts of western Nepal near Khaptad National Park and in Bhutan and Arunachal Pradesh, the climate supports a rich and diverse forest cover. Various forest surveys that have been completed in the middle-mountains zone show how the distribution of forests follows the lines of both latitude and longitude and the abrupt gradients of altitude. In the middle mountains of Nepal, at around 2,000 meters, chir pine (*Pinus roxburghii*) and sal (*Shorea robusta*) forests are widespread. Above that, deciduous hardwood forests dominated by oak (various *Quercus* species) are common. At elevations above 3,000 meters, alpine conifers dominate.

Conditions of temperature and humidity, which partly determine the distribution of forest types as well as the potential for farming, vary widely across the zone. In general, they both decrease toward the northwestern portion of the range. In consequence, the middle mountains in Kashmir and Garhwal tend to be colder and drier than they are in Nepal or in places located farther east in Bhutan and north of the Assam Valley. Pine, spruce, and cedar trees are the important components of the temperate forests in the northwestern region. In the far eastern portion of the middle-mountains zone—in Sikkim, Bhutan, and Arunachal Pradesh—the climate is much wetter and warmer. The floristic diversity is very high there. Extensive rhododendron forests, populated by numerous local ferns, bamboos, and epiphytes, give the eastern middle-mountains zone an especially lush look.

The middle mountains throughout are densely populated by human settlements. Farming communities spread across the lower slopes and dot the river valleys. The agricultural practices of the villagers have cleared much of the natural vegetation around the settlements. The largest areas of intact forests occur mainly along remote ridgetops and on very steep slopes. Smaller forest patches also reflect various cultural practices, including community forests, farmland reclamation, and sacred groves near the monasteries and temples. Elsewhere, the farmers have carved huge sections of some mountains into flat terraces in order to make room for their fields. The beveled surfaces, watered and planted in food grains, cascade in a stepwise fashion down the steep slopes, sometimes encompassing several kilometers of vertical relief. These terraces make a compelling sight and show the care with which people traditionally manage the land. Fodder for livestock and fuel wood for cooking and heating homes are gathered from the village forests. Streams are diverted in elaborate, handmade canals to irrigate the fields. Natural springs provide drinking water. Walking trails connect the villages, and everywhere people are on the move.

The settled landscapes of the middle-mountains zone thus show the remarkable imprint of human design and ingenuity. They have sustained human life for many generations, but high population growth over the past several decades, coupled with the commercial demands for alpine resources, have lent a precarious nature to the middle-mountains ecosystems. Human economies have in some places destabilized entire landscapes, making the natural ecosystems especially vulnerable to environmental change. Such events typify, for example, the middle-mountains zone in Nepal, which is the main setting for the theory of Himalayan environmental degradation, but similar circumstances are found all across the range.

It would be remiss, however, to outline the environmental damages in the middle mountains but to ignore the efforts of mountain communities to resolve them. Such a picture would be incomplete and misleading. The reclamation of eroded slope lands, the careful husbandry of soil, the regulation of forest use, the maintenance of protected forest groves, and the judicious allocation of water for irrigation are strong components of the indigenous systems of environmental management. They are found throughout the densely populated middle mountains and show the benefits of long-standing cultural practices. The deep knowledge about the local natural history registered in the cultural landscape counters the impression of imminent environmental collapse. Meanwhile, government-sponsored tree nurseries and planting schemes are new additions to the middle-mountains landscapes. They facilitate environmental restoration, which is established now on many degraded lands. Pasture conservation is also widely practiced nowadays. Livestock commonly are stall fed, instead of being taken into the forests to graze. That commonly results in less pressure on the forest understory. Such trends show signs of a slowly reconstituted balance between nature and society in many places in the middle-mountains zone.

Despite these positive trends, several problems continue to plague the zone. The deterioration of farmlands and the scarcity of natural resources compel people to leave the zone in search of livelihoods elsewhere. As a result, recent demographic trends include a gradual slowing of population growth in some parts of the middle mountains, notwithstanding the continued growth overall in the Himalayan region. In Nepal, for example, the proportion of people living in the middle mountains dropped from 60 percent in the 1960s to about 45 percent in the 1990s. Similar trends are noted in the Garhwal and Kumaun regions of the Indian Himalaya, where emigration also is common. The reports of a stabilized population would be heartening were it not for the fact that it is a result of destitute people leaving their mountain homes because they simply cannot survive in them any longer. The migrations also mean that the adjoining areas, notably

MOUNTAIN PROFILES

Terraced landscapes of the middle mountains in Nepal show the careful management of slopes and soil by local farmers. The terrace steps, which provide the only flat surfaces for agriculture, require constant attention to safeguard them against erosion. Photo by P. P. Karan

the piedmont plain and the Himalayan towns, must accommodate even more people.

THE OUTER FOOTHILLS AND THE PIEDMONT PLAIN

The southward extension of the Himalayan range composes a zone of low hills and rolling plains that averages less than 1,500 meters in elevation. The low ridge lines follow the approximate alignment of the Himalayan system. Numerous broad longitudinal depressions, called *dun* valleys, intervene among the foothills. They make good farming areas. In India, the foothills zone is known as the Siwaliks, and in Nepal it is called the Churia Hills. Viewed from the south, the hills show as a series of low green rises in the land that eventually dissolve against the Great Himalaya background.

The outer hills are composed of detrital earth material, such as clays, sandstones, limestones, and conglomerates. They are underlain by broad anticlines (upfolds) and synclines (downfolds) in the bedrock, which give an undulating look to the land. Pebbly surfaces and large amounts of loose earth material show the origins of the hills to be partly in the weathering of the high mountains. The rivers have carried the sediments from the eroded highlands and deposited them among the foothills. The largest of the dun valleys, filled with rich alluvium deposited by the Himalayan rivers, are some of the most productive farming areas in the entire range.

Significant forest areas still remain in the foothills zone, despite recent years of accelerated timber cutting. The lowland forests provide habitat sanctuaries for the region's endangered fauna. Wildlife sanctuaries at Jim Corbett National Park in India; at Chitwan, Royal Bardia, and Koshi Tappu national reserves in Nepal; and at Royal Manas National Park in Bhutan are examples of protected forest areas aimed at conserving some of the last remaining wildlife habitat in the Himalaya.

Between the outer foothills zone and the Ganges lowlands is the piedmont plain. In Bhutan, where it is known as the Duars Plain, the piedmont constitutes a sediment-filled tectonic basin formed initially by the folding of Tethys strata and later by the drainage of rivers. On the tarai in Nepal it is overlain also by river sediments. All across the Himalaya, the piedmont plain is susceptible to flooding from the mountain rivers during the summer monsoon rains. The annual

The tarai region of Nepal is an area of low foothills and intervening valleys. It provides some of the most productive farmland in the country. Photo by P. P. Karan

floods replenish soil fertility but also produce unpredictable river conditions and cause problems for human constructions such as houses, bridges, roads, and dams.

With the control of malaria in the Himalayan lowlands in the 1950s, the piedmont plain became an important settlement area for immigrant farmers. It now is the fastest-growing, most densely populated environmental zone in the entire region. In addition to extensive land clearing for agriculture, the zone's forests have been heavily cut by timber concessionaires, reducing further the area of natural habitat for wildlife. New industries and commercial activities extend now across the foothills and plains, concentrating in the market towns that have sprung up there during the past few decades.

In Bhutan, much of the new industrial expansion proposed in the country's development plans will occur in the Duars Plain along the kingdom's southern frontier. The country's principal new urban settlements are also located there, alongside the new roads. Phuntsholing, with a population of over 30,000, is an important new commercial and industrial link between Bhutan's mountainous interior and the industrial plains of India. In Nepal, many towns in the tarai grew quickly with the completion in 1982 of major sections of the 1,055-kilometer East-West Highway. The ongoing construction of numerous feeder roads into the middle mountains from the East-West Highway has created additional urban cores near the road intersections. Among the major towns in the Nepalese tarai are Nepalganj, Birganj,

Janakpur, and Biratnagar, all of them very fast-growing places. They link both southward to the Indian border and northward to the service centers newly located in Nepal's middle-mountains zone. The tarai towns now compose almost one-half of Nepal's total urban population. The high level of rural to urban migration in Nepal predicts the continued growth of the tarai towns at rates between 5 and 10 percent per year.

To the west of Nepal, in the lowland regions of outer Garhwal, the railheads and roadways at Dehra Dun, Haridwar, and Chandigarh give those towns a special prominence as gateways into the Himalayan hinterlands of northern India. Much of the new industrial activity in the Indian mountains is concentrated there. The buildup of infrastructures and urban centers across the piedmont plain allows the extension of lowland interests into the remote highlands. In that way, the new entrepôts constitute a regional development strategy that links the principal administrative and market centers with the rural highland areas. The new roads and towns in the zone provide especially convenient routes into the mountains for the purpose of natural-resource extraction. Hence, the zone is situated strategically for the future commercial development of the mountains.

The various landscapes described above are linked also to diverse modes of cultural adaptation in the Himalaya. The most popular cultural models propose a rather neat scheme whereby people exploit the environmental zones in predictable ways. Where and how people farm, live, or raise livestock are linked in the models to prevailing environmental conditions. Although they tend to be overly deterministic, these models do manage to describe the general conditions of life that still obtain in many places. But by failing to accommodate the new political-economy factors, as well as changing environmental circumstances, the cultural models prove inadequate to fully account for life in the contemporary Himalaya.

Some scholars have proposed the idea that the advancement of roads into formerly inaccessible places opens up entirely new possibilities for mountain people and thereby lifts the vertical controls on the mountain economy. The introduction of new crop types and new markets that, along with the placement of irrigation and hydroelectric facilities, accompany the roads become the tangible extensions of lowland society into the mountains.[4] Of course, the changes in mountain geography are more complicated than simply new roads. The technology and infrastructures introduced into the mountains accompany new government policies and social institutions. These, in turn, influence the ways in which native people value and perceive land and also the possibilities of social life. The contemporary Himalayan worlds have a whole new sense of drama and a decided change in character.

The valley of the Baspa is located in Himachal Pradesh, India. It is a place that is renowned for its beauty and temperate climate. In 1994 the valley was opened for the first time to visits by outsiders. Photo by D. Zurick

The Shape of Land and Life

THE HIMALAYAN DILEMMA contains a rhetoric that confounds the
policy makers; its daily reality, meanwhile, touches upon the lives of
almost everyone who lives there. To accurately assess nature and society change in the mountains, it is necessary to consider the manner by
which villagers deal with the problems that confront their highland
worlds. It takes more time now for people to collect the fuel wood to
heat their homes, the water to drink, and the fodder to feed their livestock. The productivity of many of the mountain farmlands is diminished in the face of soil erosion. Landholdings are made less secure by
degradation and political edict. Such things were best secured in the
past by direct experience, careful attention, and intuition. In modern
times, though, new ways of life demand new experience and therefore
may be poorly understood. The villagers may have little authority over
the advances of modernity that will change forever the circumstances

3

of their lives. The rapidity with which the Himalayan societies must accommodate the demands of the wider world is astounding, far surpassing any previous transitions.

The newly built roads, for example, have opened what had been inaccessible terrain to vehicular traffic and to new economic markets all over the range. Since the completion in 1962 of its first road from the southern border town of Phuntsholing to the national airport at Paro, Bhutan has built almost 2,500 additional kilometers of roads. Neighboring Sikkim, meanwhile, has trebled its roadways since 1954. Nepal began building its roads outside the Kathmandu Valley in 1953; forty-five years later, the country has 7,330 kilometers of roads. In 1992, India for the first time opened its military roads in the western Himalaya to public traffic. Such efforts both materially and symbolically convey to the remote villages the opportunities and risks of the wider world. The route to Shangri La is now paved.

Hydroelectric projects have become a common form of energy development in the Himalaya. They tap into the tremendous reservoir of power contained in the flow of the mountain rivers. Their turbines generate electricity for factories and for export to the industrial plains. They illuminate the village homes and bring to mountain people the dubious benefits of satellite link-ups. The costs of such innovations may be high. The hydropower schemes range from the giant 226-meter-high Bhakra Dam located on the lower Sutlej River near Bilaspur to the tiny run-of-the-river systems found in villages throughout the Himalayan region. When it was completed in 1963, the Bhakra Dam reservoir flooded 17,864 hectares of land and 371 villages, forever displacing 36,000 people. The loss of the traditional landscape to the impounded lake is still visible during the dry months, when the lowered waters in the Govind Sagar expose the stained white tops of numerous Hindu temples marking the position of the old town that is now submerged.

A short distance to the east of the Bhakra Dam, in the Uttarkhand Hills of India, the villagers have been struggling for decades to halt the construction of the Tehri Dam, preferring to keep their temples in clear view. They correctly believe that when the dam is built it will inundate their valley, flooding farmland, villages, and forests. The Tehri Dam would be one of the world's largest hydropower schemes, and its site sits squarely atop one of the most active seismic zones in the entire Himalaya. The obvious concern of many people is the safety of such a massive installation.

Meanwhile, sixty run-of-the-river power schemes and thirty major reservoirs are proposed in neighboring Nepal. The seven largest of those—including the hotly disputed Arun III project, located in a densely populated area in the eastern sector of the kingdom—would inundate 119,000 hectares of land and displace 120,000 people. In Sikkim, a current proposal to build a thirty-megawatt run-of-the-river scheme along the Rathong River below Mount Kanchenzonga has met with sharp opposition from the normally reticent tribal people who live in the area. The Lepchas believe the region is sacred land and therefore off-limits to such developments. That view of the world, of course, clashes sharply with the technical proposals offered by the dam engineers. The overall potential for hydropower development exists all across the Himalayan range. It is, in fact, one of the region's most lucrative prospects. But to harness the power of the rivers will lead to a great loss of cultural autonomy and to a violation of indigenous land rights. The inundation of farms and forests by the dammed waters will diminish the natural productivity of the mountain ecosystems. In the minds of many national planners, those costs are necessary in order to electrify the mountains and generate wealth for national economies. For the villagers, though, they may prove devastating.

The development of roads and dams clearly uproots people's lives. Economists argue, however, that the overall benefits of the infrastructures far surpass their costs. Ostensibly, road and dams foster economic growth in the mountain villages. That, at least, is the most common argument used to promote them. Too often, though, they merely extract the wealth from the mountains and send it to the Indian plains or to the urban centers, leaving the villagers stranded. The roads allow the rapid removal of mineral, timber, and other natural resources, while the hydropower plants harness water to make electricity to be sold downstream. In either case, the flow of wealth follows the pattern of infrastructure, ultimately concentrating in places outside the Himalaya. The social and environmental costs, meanwhile, stay in the mountains, borne by the people who live there. That is the irony, and inequity, of such developments.

But more imaginative forms of change also exist. These projects propose alternative ways of managing economic development in order to minimize its adverse impacts and to make it more compatible with local circumstances. The establishment of parklands and protected areas, such as the Sagarmatha National Park in Nepal, the Great Himalaya National Park in India, and the Royal Manas National Park along the Bhutan-India border, introduce new economic programs that may actually enhance local landscapes and cultures. Cottage industries based upon the manufacturing of handicraft items, beekeeping, herb gathering, and the commercial development of other forest products are common examples of village-based activities that may provide valuable income to people. Nature tourism, when properly managed, also offers additional economic possibilities. The fact that villagers seek to engage in such economies shows that they are not bound to a romantic ideal of tradition, but rather to a careful balance of economy, culture, and environment.

Such local programs are not without their faults. They are criticized because their maturation time is slow. They may never assume the huge presence of the big development schemes. The smaller programs by design seek to avoid the concentration of wealth and power that commonly accompanies large-scale infrastructure programs. That makes them less influential in national circles. Nonetheless, the alternative schemes already show some new and locally responsible pathways for change. The Annapurna Conservation Area Project in Nepal has become a model for similar conservation efforts around the world. Where the environment programs work best, they enlist the cooperation of local people in their design and implementation and show how they mean to serve above all the needs of the local communities rather than some distant national purpose. In light of the alienating forces of modernity, where regional goals may be at odds with local affairs, such alternatives are reassuring to many people.

Nonetheless, the new roads, factories, and dams remain the most visible signposts of the future. They are ubiquitous across the Himalaya. But the economic policies that guide them may stay hidden from the view of most people. That is because they emanate from places far removed from villages. The desire to improve the economies in the mountains remains a necessary effort. The per capita annual incomes in the Himalayan region are among the lowest in the world, averaging about $160. But the big programs depend heavily on foreign assistance for their financing, taking the Himalayan countries deeper into debt, and they tend to favor lowland populations, not the mountain communities. In consequence, the region's meager revenue tends to concentrate in the cities and towns, such as Gangtok, Kathmandu, or Shimla, and among the offices of government, foreign consultancies, and businesses located there. Big schemes often disregard the natural constraints of the fragile mountain ecology and thereby contribute to many of the problems they are meant to fix.

The alternative programs that seek to minimize the pressures of economic activity and demographic change on the landscape themselves now augment planning programs throughout the Himalaya.[1] They are cause for some optimism. The International Center for Integrated Mountain Development (ICIMOD) was established in Kathmandu in 1983 as a regional research and training center with the explicit mandate of sustainable development.[2] Toward that end, ICIMOD sponsors programs in eight Himalayan countries and serves as a regional center of information and expertise on such topics as pasture management, crop and irrigation schemes, forest-products use, ecotourism, and regional planning. Despite some criticism leveled against ICIMOD for bureaucratic inefficacy, the international scope of its programs provides some of the most innovative development planning in the Himalayan region. The sustainable-development model promulgated by ICIMOD has become the cornerstone of Bhutan's economic policies. Lyonpo Dorji, the chairman of Bhutan's planning commission, wrote in the preface to the country's *Seventh Five-Year Plan (1992–1997)*, "The focus of development in Bhutan is not only economic growth or material prosperity for its people . . . Our national aspirations also include improvements in the quality of people's lives . . . preserving Bhutan's special culture and maintaining our precious natural environment. We are also concerned to ensure that progress be spread as equitably as possible in all parts of the country."[3]

Such national programs as Bhutan's, supported in part by the activities of ICIMOD, seek to put the Himalayan economies on a more sound footing by emphasizing local participation and fiscal responsibility in small projects rather than by accepting the less manageable, much bigger projects promoted by many international donors. They position local communities closer to the center of decision making and emphasize local needs over commercial resource extractions.

The village of Nakho, situated along the India-Tibet border, shows the traditional Tibetan style of settlement. Cloistered homes made of sun-baked mud surround a small impounded lake. Agricultural fields and pastureland are located nearby. Photo by D. Zurick

The Himalaya, though, is still viewed by much of southern Asian society as a peripheral place. Were it not for their great resource potential, strategic military significance, or recreational opportunities, the mountains might be seen to lie fully outside the concerns of the plains society. That perception denies the history of the mountains, though; the Himalaya contain enduring places where people have lived for centuries. The abiding relations between nature and human society have forged strong local territories in the mountains. Such places are not merely mirrors of global change or the picturesque backgrounds to grand social theories but rather full participants in global transformations.[4] The Himalayan region is precisely at that point in its history when local cultures and landscapes, confronting head-on the demands of the global economy, must reconcile world events with their own aspirations and environmental capacities. How that plays out will be largely determined by how well the changes are understood.

Humans and Nature

Ethnographic studies conducted in the Himalaya by geographers and anthropologists, beginning only in the early 1960s, tell us a great deal about the farming methods of the villagers, the seasonal movements of the alpine pastoralists, and the exchanges of the caravan traders. We know something about the complex rituals of the Buddhist and Hindu peoples and how they regulate the human use of the land. We understand how family practices and reproduction may organize village society and resource management. Those studies all show the power of culture in determining mountain life. As a result of the past forty years of study, we have a somewhat better idea also about how local communities may change as they link with the larger regional systems and how the natural environment may be transformed in the process. We can measure current events against mountain life as it was lived in the near past.[5] By looking into the archival record, we may gain an even longer look at the changes in society and nature, dating back for several centuries. That historical perspective is invaluable when considering the current, frantic pace of change in the mountains.

Numerous ecological themes dominate the framework of Himalayan scholarship.[6] They include resource extractions by mountain people, the practices of agriculture, the emergence of territorial politics, and how the factors of culture and economy influence the ways in which native people interact with the natural world. Such practices are rooted in a long stretch of time and in a local place. In a modern sense, though, the connection between villages and the wider world has come to dominate the rural scene. To comprehend that, it is necessary to look at how the villages both maintain their long-standing traditions and accommodate the interventions of outsiders.

The study of land and life in the Himalaya constitutes a dialectic, or a two-way discussion, in which the natural and social environments in a very real sense *explain each other.* Such an understanding is based on the notion that in social terms, nature and human beings are interdependent and mutually reinforcing to the degree that the existence of the one actually depends upon the other. That idea has concrete expressions. For example, the management of environmental resources—one of the main components in the Himalayan "ecocrisis" scenario—depends upon their physical availability. That is determined in part by local natural capacities such as water, vegetation, or soil, but it also is a matter of human interpretation: the recognition that a natural product has a potential use for human society. That realization, in turn, is mitigated by land ownership and resource entitlements that give people authority over the use of natural goods.

The control of resources in the Himalaya is tied to politics, wealth, and social status. People may own land outright or live as members of a community who enjoy shared rights to the land. The latter case is especially true among the mountain grazing communities. The forests, in turn, may be state property, held in common by the village, or owned outright by individuals. A forest's status will influence how it is actually used. The nexus of resource use in the Himalaya occurs where the natural systems and the systems of indigenous knowledge and regulation meet the larger political economy. In order to comprehend environmental trends in the region, then, it is necessary to explore the forces of contemporary society at work there.

Ecological threats to the mountain landscapes constitute part of a worldwide problem of land degradation. The very fabric of the planet is being torn asunder, compelling an international study of landcover change that includes the Himalayan region. The loss of the unique biological heritage of the mountains puts the entire earth at further peril. But the issues of global change and biodiversity must also translate to the immediate needs of villagers and to concerns about the dwindling environmental security of native people. Under the conditions of risky land tenure, such as when new government curtailments are put on land use, the incidence of land degradation tends to be more common. That became evident in Nepal when the village forests were nationalized in the late 1950s. The new government ownership of the forests usurped community systems of forest use, replacing indigenous controls with cumbersome new national forest regulations. The change in the rules led to the destruction of forest areas because local people were no longer in charge of protecting them. Similar circumstances occurred when the new Himalayan parklands were established. The native people lost much of their traditional access to the forests and pastures located in the park territories, even though they had been using them for many generations. The result has been a common disregard of the park rules about fodder and fuel-wood collection. In many cases, the people who lose the most under the new environmental guidelines are the very poor people, for whom few options exist.

The current ecological crisis in the Himalaya is foremost a crisis of poverty; poverty induces environmental damage and is deepened by it. Conflicts over mountain territory—a problem that dates back to the region's colonial history, as well as to modern nationalism—are framed in the issues of poverty and cultural survival. Numerous and competing views on the land, urging incompatible systems of land use, exist among the native residents, the developers, and the government officials. For example, the commercial development of mountain landscapes for the purpose of orchards and market grains, a common occurrence where roads exist, undermines the traditional farm villages' cultural authority, which safeguards people against the perfidy of the export economy and provides much needed support during hard times. The environmental challenges that will take the Himalaya into the twenty-first century spill across the concerns of the villagers, the national economies, and the global community. Social and environmental changes are linked to events that operate at various geographical scales.[7] At the international level, the territorial aggressions of some Himalayan countries have led to boundary disputes that influence how land and resources are used. At village levels, environmental conflicts emerge as a result of competing economies.

Historically, the mountains were organized under the colonial aegis of the British, Russians, and Chinese, who sought to bolster their respective imperial claims. Local conflicts, meanwhile, centered on feudal properties, village common lands, crown land, and inherited landholdings. The postcolonial nationalism in the Himalaya has imposed a new set of claims on the mountain places, initiating in turn a period of strategic boundary movements in the region. In some cases, that has led to international warfare; for example, between India and Pakistan in Kashmir and between India and China over the Aksai Chin Desert north of Ladakh. The separatist struggles in Kashmir and Ut-

Rapid change has followed the road into the Ziro Valley, Arunachal Pradesh, India.
Photo by P. P. Karan

tarkhand, the refugee settlements in Bhutan and Nepal, and the insurgencies in the far eastern Himalaya all raise additional questions about the nature of territorial conflicts and the role of local sovereignty over land resources.

Knowing how mountain territories are organized at various levels of political control is crucial in understanding how the land will be managed. The native peoples' views of the mountains, and their local resource strategies, are often measured against national and international development programs. The latter have the support of international finances and include the establishment of horticulture and other forms of commercial agriculture, the exploitation of forests for timber, the development of tourism, and the harnessing of river power for electricity. These developments describe the new forms of land use in much of the Pir Panjal range, along the valleys of the Beas and Sutlej rivers in Himachal Pradesh, in Nepal, and in many other parts of the range. They compete with the subsistence economy for rights to the land, resulting in expanding rural industries, which is their intention, but also in degraded natural environments. Mountain conservation efforts, meanwhile, include the various Himalayan parklands and tourism programs. They propose that the economic programs be combined with conservation goals. Good examples of such efforts are found in the Annapurna and Makalu-Barun areas in Nepal, where important new conservation parks have been initiated. Other Himalayan localities propose similar designs, including the Great Himalaya National Park in India and the Black Mountains Nature Reserve in Bhutan.

A tenuous balance exists in the Himalaya between the security of tradition and the risks of modernity, the desire for economic growth and the need for environmental conservation. These dualities give rise to a certain tension in the mountains and position the villages differently, not as quaint and removed backdrops in the play of regional politics but as full partners in Himalayan affairs.[8] Despite their striking appearances, the mountain landscapes are vulnerable places. In part that is a natural condition set by their fragile ecology and active geology,[9] but many areas also exhibit pressures on the land due to growing populations and insecure land ownership. That leads to higher levels of resource exploitation for both subsistence and commercial purposes. Moreover, the development of recreational pastimes in the mountains brings hordes of international tourists to the region.[10] They have their own impact on the land and cultures of the range. The growing demands on the mountains have the potential to undermine the sustainability of villages and to irrevocably damage the mountain ecosystems. As a result, they all have become the concern of conservationists working in the region.

The biological diversity of the Himalaya, worth preserving in its own right as well as being a global issue, exists in tandem with the region's great cultural diversity. The two are in fact interdependent. According to Anil Aggarwal of the Center for Science and Environment in New Delhi, "Cultural diversity is not an historical accident. It is the direct outcome of the local people learning to live in harmony with the region's extraordinary biological diversity."[11] With interested parties holding that in mind, the global and regional efforts to preserve the Himalayan environment can be accomplished best by attending to its cultural needs. This viewpoint is supported by the fact that the most successful efforts to preserve natural ecosystems in the Himalaya have all begun by viewing them as places where people live.

The diverse mix of people who inhabit the range brings to the Himalaya a rich cultural heritage. Its traditions come from four major human realms—the Islamic traditions of northwestern Pakistan and Kashmir, the Hindu cultures found in Garhwal and Kumaun and throughout much of lower Nepal, the Tibetan monasticism in the trans-Himalaya borderlands and in Bhutan, and the tribal life of parts of eastern Nepal and Assam. Because all those groups actually intermingle freely, broad transitional areas occur where cultural traits are commonly borrowed and shared.[12] Nevertheless, a few relatively dominant types of economy prevail in the mountains.

Nomadic pastoralism, which has people and livestock moving more or less continually throughout the year in search of grass and water, is found only in the Tibetan plateau zone and among the pasture lands of the Great Himalaya zone. Relatively few people belong to the nomadic communities. A group of people who practice a related lifestyle are known as agropastoralists. They combine farming in settled villages with summer migrations to the highland grazing areas. The agropastoral communities are dominant in the more protected valleys of the Great Himalaya zone and among the higher elevations of the middle-mountains zone. Sedentary farming, based on growing wheat, corn, or rice, is found all across the lower ranges. It supports the largest concentrations of people living in the mountains. The various economies, mixed with a few lesser types, such as shifting cultivation and caravan trading, traditionally determine how mountain dwellers exploit the multitiered zones of their natural environment. But the new commercial developments that have taken hold in many areas compete nowadays with the traditional adaptations for land and labor. They predict the emergence of totally new lifestyles in the mountains.

The traditional farming scene in the Himalaya is a complex one. Types of landholding are distributed widely across a range of microenvironments. Farmers rotate their planting times and mix together a number of hardy indigenous crops to help minimize the risks of such hazards as frost and pests. The annual migrations of people and livestock from the settled farming villages into the high pastures takes advantage of both privately owned village land and the commonly held grazing areas. Pastures are regulated according to long-standing village traditions. The grains that are harvested in the village fields, the dairy products obtained from the herders, and the other food items produced or gathered in the forests are traded among the mountain communities in order to insure sufficient resources for everyone.

A general model of mountain culture called *Alpwirtschaft* commonly appears in the Himalayan literature to describe the mixed mountain economy. "Alpwirtschaft" is a German term, used initially to describe the economy of the European Alps, that refers to a combination of farm and pasture landscapes. People move seasonally between the two, utilizing the high mountain grazing lands during the summer months and residing the rest of the year in the permanent villages at lower elevations. The Alpwirtschaft model, with its emphasis on resource management and community sharing, has provided scholars and policy makers with a fairly good basis of comparison for the traditions of people living in the remote areas of the Himalaya,[13] but nowadays the affairs of villagers must also conform to new patterns of political and economic development. The enhanced accessibility of the mountains by way of roads, airports, and telecommunications, for example, is a major factor in transforming society and land use throughout the range. These latter bring the global economy to the mountain thresholds, rearrange the landscape, and position the villagers only a step away from a new world order.

The agricultural innovations that accompany the infrastructure alter farmers' production strategies, increasing yields for some but making everyone more dependent upon costly machinery, fertilizers, pesticides, and seeds. As a result, the farming communities tend to rely more on the national and global economies for their livelihoods. As the cost of farming goes up, so too does the potential earnings of those farmers having sufficient assets to invest in the new commercial opportunities, but marginal farmers may not be able to participate in the benefits that come from the development of mountain resources and may become even more impoverished.

The new travel routes into the mountains and the development of recreational industries alongside them further define life in the contemporary Himalaya. Tourism shapes new ways of managing the environment for economic gain, ways that may reduce the environmental security of mountain people. It may even alter the values that people place on the natural world and forge new associations between them and the places where they live. In the extreme, the mountains may become a theme park, where the tourists play and the villagers watch.

In this context of sweeping transformations, in some cases it no longer is clear who the land managers are. The sustainable-development programs insist that the local communities should keep control over local resources; however, many development programs do not adhere to this policy. The decisions to develop mountain resources for industrial uses are most commonly made in political centers that are far removed from the lives of mountain villagers. The old view of mountain life holds that the Himalaya is a region of isolation, of tightly proscribed land uses, and of static societies, but that view does not capture very well the dynamics of nature and society in the modern mountains. For that, we must look past the old environmental models and toward the new regional and international alignments.

New Age Alignments

The agencies of change in the Himalaya generally move up the river valleys to form longitudinal corridors. That is the most convenient route for roads, markets, and other developments. They cut across the grain of the mountains, not along the altitudinal contours that determine much of traditional life and geography. The valleys of the Beas River in Kulu and the Kali Gandaki River in Nepal are outstanding examples of such corridors. Intensive change has occurred in both these places in the past two decades. Kulu has become one of India's premier horticulture and tourism areas. Orchards, hotels, and billboards are now ubiquitous features in the landscape. The Kali Gandaki valley, meanwhile, is an old caravan route that now literally bristles with commercial villages, orchards, brandy distilleries, hydropower stations, and tourist chalets.

The north to south corridors result from several notable trends other than the fact that they occupy river valleys. The corridors tend to lie directly north of major urban centers in the lowlands. Where the cities are growing fast, the linkages into the mountains are greatest. The recent changes in the Indus valley, for example, are tied to the construction of the Karakoram Highway, which connects the Grand Trunk Road in the south with the China border. The influence of Rawalpindi and Islamabad, two of Pakistan's most important northern cities, now penetrates the farthest reaches of the Indus mountains. In Nepal, the important growth corridors extend north from the tarai towns of Dhangadhi, Nepalganj, Birganj, and Biratnagar into

The National Road in Bhutan connects the India border with the capital city of Thimphu. Photo by P. P. Karan

the middle mountains. The new feeder roads also link the mountain areas with markets located in India. Farther to the east in Sikkim, the Siliguri-Gangtok route connects highland Sikkim with the industrial centers of the Indian plains. The transportation networks in Bhutan tie Phuntsholing to Thimpu, Gaylegphug to Tongsa, and Samdrup Jonkhar to Tashigang, thus effectively connecting many of Bhutan's mountains with the southern Duars Plain. The new roads into the remote Paktai Hills of Arunachal Pradesh bring the Nochte and Wanchu tribes into direct contact with the loggers, tea planters, and government officials who reside in the hills above the Assam Valley.

The extraction of mountain resources for commercial purposes inevitably follows along such motorable routes. The roads that cross into formerly remote Kohestan encourage major new timber cutting in Pakistan's Hazara and Mansehra districts. Market towns, industrial parks, and horticultural zones straddle the Beas, Sutlej, and Bhagirathi river valleys in the mountains of Garhwal, displacing in spots the areas of forests and farmland. Hydroelectric plants and telephone poles dot the valleys in the Himalayan range where industrial growth is promoted. Pilgrims and tourists also travel along the new roads, stopping en route at new settlements and commercial attractions. Streams of job seekers move down the mountain corridors, while businesses and trade move up them.

But new spatial trends do more than merely draw lines on the maps. They shape a new geography and cut new edges into the landscape. The mountains are more than ever extensions of industrial society, even as places remain primarily rural and agrarian. The national governments tend to give most support to the activities which develop mineral, water, and forest resources. These activities create wealth in the mountains while at the same time posing serious threats to the well-being of local ecologies and peoples.

Air and water pollution, once unknown in the Himalaya, now plague many areas. North of the Doon Valley in India, within sight of the town of Dehra Dun, the mountain road to the Mussoorie hill station bypasses numerous limestone quarries that date back to the 1960s. A history of unregulated mining in the quarries has scarred the face of the mountains, contaminated the water supply in the valley, and regularly sent clouds of rock dust rolling across the valley floor. The once pristine waters of Dal Lake in Kashmir and Nainital Lake in Kumaun are heavily polluted by local industries and household waste. The levels of air pollution in Kathmandu are now among the highest in the world. On most days, the contaminated air prevents the city's residents from even seeing the mountains that surround them, and it forces many of them to wear face masks for protection. The dusty and toxic air in Kathmandu results from the absence of regulations on emissions from vehicle exhausts, from cement and brick factories, and from numerous other polluting sources located in the Kathmandu Valley. Discharges from factories located in the valley flow unchecked into the rivers. The chemical dyes from Kathmandu's numerous carpet factories transform the crystal clarity of the holy Bagmati and Vishnumati rivers into grotesque metallic red, green, and blue hues. Although industrial pollutants have not yet become widespread beyond the valley, they have made life in the city unhealthy. In the mountains widespread pesticide and fertilizer contamination occurs where commercial agriculture is practiced. The contamination is especially acute in the orchard areas of the Sutlej valley and in other intensive horticulture zones.

The adverse environmental circumstances result mainly from the laissez-faire attitudes that describe Himalayan development since the 1960s. Basically, economic growth is given priority over most other concerns. Only recently, as the ecological problems have worsened to the point of international notoriety, have new ideas about conservation become more popular in government circles. Many of the national plans in the 1990s, for example, link poverty in the Himalaya to the high levels of land degradation. The planners, then, are required to contend with the issues of environment, cultural preservation, and social equity. It is a large, still unmet, challenge.

To help envision how local communities connect with the wider social systems and to understand the consequences that may ensue, it is useful to consider how the Himalayan worlds interact. On the one hand, the *vertical* world of farm production, land tenure, and environmental relations anchors people in a particular place. It connects villages and land with long-standing sets of ecological, social, and technological arrangements. Indigenous systems of knowledge determine the manner of local environmental management and the traditional use of natural resources. Agricultural labor exchanges and resource sharing further reinforce the social relations and help to accomplish the work tasks. Community-based systems of land tenure allocate access to pastures and forests for the purposes of livestock grazing, wood-cutting, and foraging. They all regulate the sustainable use of shared resources.

The *horizontal* world of social institutions, markets, and political economy, meanwhile, extends the reach of national agencies to Himalayan households and villages.[14] Although the traditions of the past still guide people today, the villages never were static or isolated places. The old trading routes, the migrations for work, the weekly markets, the religious pilgrimages, and other cultural practices have always connected places in the mountains. The new developments now bind the villages with a much wider region, linking people, not to the land necessarily and to their pasts, but to the nation and world and perhaps to an entirely new vision of the future.

Where village life meets the world of market transactions is the fullest possible context for understanding the contemporary cultural and environmental changes. That meeting creates a tenuous position, however, and a delicate balance exists between nature and society that may be easily disrupted by the forces of modernity. To a great degree, the level of stability in Himalayan life today reflects also the history of territorial conflicts in the mountains. They have shaped the course of economy and society over time and have influenced the rights of people over land and resources. These factors are key to understanding contemporary environmental dilemmas in the region.

PART TWO

POWER AND TERRITORIAL CONQUESTS

A Hindu temple and pilgrimage site at Kedarnath in the Garhwal region of India. This is one of the most important religious sites for Hindus in the Himalaya. Photo by P. P. Karan

Myth and Prehistorical Territory

THE HIMALAYA MOUNTAINS OCCUPY an auspicious place in the worldviews of their native people. They are the southern gateway to Mount Meru, the mythical center of the Vedic universe, embodied in its earthly form as Tibet's sacred Mount Kailas. The widely worshipped Hindu deity Shiva is believed to live atop the summit of Mount Kailas, where he cavorts with his consort, Parvati, in Lake Manasarowar, which forms at its base. He is the premier mountain god of all Hindus, and occupies the pinnacle of a vast cosmology that spans the highland domain of the Indic world. The great Indus and Brahmaputra rivers, which share their headwaters near Kailas, wrap around the tectonic Himalaya and embrace the mountains in a spiritual grasp that for native peoples absolutely resonates with sacred life.

The snowfields and glaciers of the Himalaya replenish the rivers Ganga and Yamuna, whose nourishing waters support life across the

4

fertile southern plains and are home to countless familiar and protective deities. The rivers delimit a sacred map of southern Asia that shows the Himalaya to be a truly celestial realm on earth. In the Hindu worldview, the mountains are created anew with every grand cycle, or *manayuga*, of cosmic life and destruction. The time span contained in each cyclical turn of the wheel of life is epic, exceeding even the long reach of geological reckoning.

The traditions of southern Asia thus know the Himalaya to transcend the normal conventions of time and space. In Vedic thought, the mountains anchor the unstable earth and support the vast heavens. They are the *axis mundi*, a direct physical link between the secular and sacred worlds of humankind. In such an encompassing view, the entire Himalayan region is a sanctified place with a divine nature. All across the range we find cultural markers attesting to the reverence the villagers display toward their mountain homes: magnificently inscribed stone *mani* walls; prayer flags atop dangerous passes; stupas that grace the Buddhist villages; elaborately carved temples and engraved shrines dedicated to resident Hindu deities; *dharamsalas* (temple inns) where pilgrims stay; and ceremonies and propitiations that everywhere regulate human behavior in the villages.

Geographical references to the Himalaya are abundant in ancient Indian texts. The Upanishads and Sutras show the mountains to be the ancestral homelands of the Hindu people. The epic books chronicle the holy places that devotees still visit when they are on a pilgrimage, known as a *yatra*. The auspicious temples, the headwaters of holy rivers, the ancient rock inscriptions, the soaring peaks, and the religious forests all are part of the sacred landscapes of the Himalaya Mountains. Devout Hindu pilgrims journey to the religious sites at Haridwar, Badrinath, Kedarnath, or Gangotri in Garhwal, to Muktinath or Pashupatinath in Nepal, or to countless other divine places scattered across the Himalaya. Amid such places, the devotees find the northern boundary of sacred India, or *Bharatavarsha*, just as surely as the geologists uncover the deformed tectonic extension of the Indian subcontinent.

The Himalaya is also a consecrated realm in the dharmic traditions of the Tibetan people, who live mainly on the high plateau in the north. Devout Buddhist pilgrims encircle Mount Kailas, which they too consider to be sacred, with full head-to-foot prostrations. A complete circumambulation may take many months to complete and is called a *parikrama*. One complete journey can erase the bad deeds of a lifetime; ten circles can wipe out the sins of an entire earth age. Such are the divine powers of ritual devotion and of the mountains. The cold, arid highlands that surround Mount Kailas are drained by the Indus River flowing to the west and by the Yarlung Tsangpo River flowing to the east, so the Himalaya are sanctified also for the Tibetans by the holy waters of those two great purifying rivers.

Located on the far eastern bend of the Tsangpo-Brahmaputra River, in the region of Kham in Tibet—a wild, mist-enshrouded landscape, visited by only the most determined pilgrims—is the revered mountain called Kundu Dosempotrang (The All-Gathering Place of Adamantine Being). It lies in a rare and lush region that the Tibetans know as Beyul Pemako, or Hidden Land of Lotus Splendor. For the Tibetans, that little-known region is a spot of legendary power, a place mainly of hermits and holy lakes. In its very lushness, Beyul Pemako balances the stark, arid landscape of Kailas. Spanning the distance between the two remote regions is the long stretch of the Himalaya. In the sacred geography of the Tibetan Buddhists, Kailas and Kundu anchor the two ends of the spiritual world of the mountains.

The old Tibetan books tell us how the early Buddhist saints cut giant swaths in the mountains in order to drain the flood waters of ancient Tibet. The massive gorge of the Yarlung Tsangpo at Namche Barwa, where it cuts through Beyul Pemako and becomes the Brahmaputra River, is one such place. Another is the canyon at Chovar, near Kathmandu in Nepal. In the Buddhist tradition, Bodisattva Manjushri sliced the Himalaya Mountains there to empty the flooded Kathmandu Valley. All across the mountains, the local cultures have their own explanations for how the land came into existence. The mythical origins of the Himalaya Mountains not only exist in the written codes of Hindus and Buddhists but embrace also the animistic beliefs of diverse tribal peoples who have settled there. These ancient stories seek to make sense out of the Himalayan world, and to affirm the native peoples' rightful place in it.

Their myths correspond strangely to the geological records, but more importantly they frame the origins of the Himalaya Mountains in the intimate, religious and cultural histories of the people who inhabit them, and they sanctify the relationship between the mountains

A Tibetan monk on a narrow trail that leads to the Nang Pa high pass between Nepal and Tibet. Photo by P. P. Karan

and people. That affiliation has withstood the centuries of territorial conquest and demands on the land.

But, where the beliefs of the native people have shaped a pious geography, the outlooks of the successive Himalayan empires have had decidedly more secular interests at heart. The dilemmas that the Himalayan people face today result in part from an ever-widening schism between the old traditions of the spirit and the new demands of a material world. Colonial and postcolonial geopolitics have gained authority in the mountains by focusing their concerns on the attributes of the land, not those of the human spirit. With economic gain in mind, the latest developments display little interest in the ways of the past. The Himalaya is a watershed not just for geographical provinces or the affairs of nation states but also for the very worldviews that compel them. Under such circumstances, the overwhelming struggle in the Himalaya is the one by native people to maintain authority over their lives and their land.

The regional descriptions of the Himalaya presented in the early pages of this book reflect the best efforts of geographers, cartographers, and anthropologists. The mapping and naming of places in the Himalaya have historically been tied to a quest for order amid the seeming chaos of mountain land and livelihoods, and for control over what appeared to be a landscape without proper jurisdiction. They are attempts to simplify the bewildering landscapes of the mountains so that we may better understand them. But in drawing the new boundaries, the regional schemes inadvertently erase some of the native marks on the land. The long history of human settlement in the range has shaped the mountains into a mosaic of well-known and intimate places. Yet, since colonial times at least, the Himalaya has been systematically surveyed and classified in order to give its territory a rational structure. These various claims made upon it by outsiders are legitimized by the assumptions of logic and cartesian space that imbue western science and thought.

The early cartographic efforts, especially the cadastral surveys of the British, were meant to create a cohesive region, and eventually to uncover its material wealth. But for many native people, the peaks that topographically shape the mountain world delimit also the sacred realm of the high country; the villagers place religious shrines alongside the trails into the mountains in order to honor the deities that dwell therein. By thus signing the path, they mark also a journey into the mythical worlds of the mountain tribes. Those who pass among the mountain landscapes enter a moral order that may be as elusive to Western reason as the chimera of ice and snow on the mountain summits. The precise boundaries of the modern nation states are set against the organic reckonings of the native people, so it is not surprising that conflicts ensue between the two; they distinguish, after all, competing worldviews. As in much of the rest of the world, the indigenous and imperial conceptions of Himalayan geography conflict with each other when it comes to such matters as territorial sovereignty and environmental management.

We can distinguish five broad historical phases that have shaped the manner of control over the Himalaya Mountain landscapes: the prehistoric period of indigenous land settlement coupled with cultural invasions and early political conquests; the medieval period of early state formation under the feudal arrangements of the Rajput princes from India; the colonial spheres of influence; the era of postcolonial nationalism; and the modern period of national development and native land struggles. These historical phases to some degree overlap, and their timings vary across the region. Nonetheless, they give to the Himalaya a comprehensive history that has forged distinctive relations between power, society, and land. The political history has laid great claims on local cultural autonomy and on sustainable land-use and resource management. In order to explicate the nature of the contemporary environmental dilemmas in the mountains, it is first necessary to unravel the territorial conflicts that have historically determined the arrangement of nature and human society there.

The Early Settlers

Little is known about the earliest inhabitants of the Himalaya. The archaeological record is based on only a few excavated sites and scattered rock inscriptions, and provides few clues about the daily activities of the Paleolithic or early Neolithic peoples in the mountains, or their impacts on the natural environment. Most of what we know about Himalayan prehistory is derived from the Sanskrit epic literature, from unsubstantiated oral tradition, or from the historical interpretations of the early foreigners residing or traveling in the region.

Records do show that the mountains served as a place of refuge for people fleeing the control of petty rulers and chieftains based in the southern lowlands. These early cultures were nomadic, moving in response to the availability of game and food plants, but their long-term survival rested on their abilities to safeguard access to forests, horticulture plots, and water. The dispersed tribes maintained territorial rights for centuries, until their gradual subjugation by the Himalayan kings and princes. The petty dynasties that arose in medieval times led to the formation of highly centralized societies and to elaborate systems of production, resource extraction, and tribute.

Prior to the emergence of the Himalayan empires, a lengthy period of spontaneous land colonization took place, during which the settlers established scattered villages based upon plantings, foraging, and animal husbandry. These early settlements remain obscure, but some broad patterns are discernible. The western Himalaya is known to have been first inhabited in the middle centuries of the second millennium B.C. when armies of warriors arrived from the northwest as part of the Aryan invasion of southern Asia. The warrior tribes conquered much of the native population when they established their residences. Most accounts of the prehistory of the western range start from that rather inauspicious beginning.

The Aryan invaders encountered groups of aboriginal people who had migrated to the mountains much earlier, from the southern plains. The aborigines had enjoyed a great deal of mobility in the mountains, free of the incessant demands of controlling rulers. The etymological links between the old tribal names in the western Himalaya and those of southern and central India indicate possible Dravidian origins among the earliest people of the mountains. Tribal people in Chamba and Kangra, for example, are called Koli, and are thought to be of the same group of people as the Kols of central India.[1] This ethnohistorical thesis is not very strong, however, and without corroborative evidence from archaeology, geographical ties between the native highland settlers and the plains people of the Deccan Plateau are mainly conjectural.

These earliest inhabitants of the Himalaya are one of the major stocks contributing to the ethnicity of the contemporary *pahari* people (meaning "of the mountains"). They are linked to the *dom* (mountain-based artisans) castes found today throughout the western ranges. A second early lineage is composed of the descendants of the invading Aryans, who are called the *khasa*. Despite considerable intermingling over the millennia, the khasa remain high caste while the dom still occupy the lower castes of the contemporary pahari population. The anthropologist Gerald Berreman speculated that at one time the khasa and the dom were two separate and homogeneous groups, the former being the superior agricultural people while the latter occupied artisan roles of servitude, but, if this were the case at one time, the centuries of contact between the two groups have greatly diminished the differences between them.[2]

The history of the doms is mainly inferred from their present-day social status. They are the untouchable groups of the Himalaya. Early commentators reported that the doms had been from ancient times the slaves of the khasa. Today the dom have no distinctive language or religion; their cultural heritage merges with that of the khasa. Whether they came from the Indian plains or whether the khasa brought the doms with them as slaves remains uncertain. The genetic mingling of the paharis over the millennia has erased any physical basis from which the geographical origins of the dom might be traced.

The Sanskrit literature, meanwhile, contains explicit reference to the khasa, who appear frequently among the Indian epic texts, in the ancient *Puranas*, for example, and in the *Mahabharata*. The khasa spoke an Aryan language, were warlike, and kept slaves. It is probable that they occupied an important position in central Asia before entering the Himalayan range between 1500 and 1000 B.C. They came from the northwest, possibly from Tibet through Garhwal, but scholars do not agree on the exact migration route, or even whether there was a large-scale invasion at all. Whatever the case, by the sixth century B.C. the khasa had settled widely in Kashmir and Garhwal, and rock inscriptions in the Karnali region show that by the twelfth century they were living in Nepal as well.

Patchy evidence shows that the khasa tribes have inhabited the middle mountains of the Himalaya for a very long time. When people from the Indian plains migrated into the hills, beginning in the seventh century A.D., a long process of acculturation occurred, during which the khasa people adopted many of the practices of southern Hindus. They took the names and status of Brahmans and Rajputs and assimilated an unorthodox blend of Hindu social and religious traits. There is today no pan-khasa identity. A great deal of regional heterogeneity exists among the pahari people of the western mountains; they speak various dialects and have different customs. The high-caste khasa, however, still dominate the low-caste doms as paternalistic overseers. Most important and to the point, the high castes control the mountain land, which means they control those who are dependent upon it.

The early migrating khasa tribes often encountered large-scale systems of political control, and conflict often arose between the khasa and the preexisting culture. In western Nepal, for example, the Bheri-Karnali region was under the rule of an autonomous Magar king as early as the eighth century A.D. That kingdom was usurped when the khasas established their political center in Jumla.[3] The Khasa-Malla kingdom, as it is known, flourished in western Nepal during the Middle Ages, until it was absorbed in the fifteenth century by the Rajput kingdoms that had become established in the mountains. As Hindu immigrants settled in Nepal, the khasa tribes adopted the Hindu forms of culture. They also intermingled with the bhotiya people, who had come into the Karnali region from Tibet during the seventh century and produced a syncretic culture that survives still in the remote regions of Humla, Karnali, and western Dolpo. Unique cultural traits of the region are evidenced today by the frequent intermarrying between castes and the practice of polyandry, by the relaxed proscriptions on meat eating and liquor drinking, and by the participation of non-Brahman castes in ritual and religious duties.

It remains difficult to say exactly what land tenure was like under khasa rule. Social and economic relations, however, were based upon custom and tradition. Villages tilled the land as self-sufficient economic units, sometimes practicing the hereditary transfer of land but paying tributary payments demanded by their khasa overlords.[4] Population pressure on the land was generally absent, and extensive forms of agriculture, including nomadic slash-and-burn techniques as well as foraging and hunting, were common. The conditions of enslavement and taxation that prevailed across the Himalaya where khasa tribes dominated were less apparent in the case of Nepal.

Early settlement in the far eastern portions of the range, meanwhile, is even more obscure than that of the western regions. The earliest inhabitants were foragers and shifting cultivators who came from the Assam hills and the Brahmaputra valley. They migrated into the mountains in advance of the expanding Han empire of southern China during the first millennium B.C.[5] Numerous cultural traits are shared among the tribal groups who inhabit the eastern mountains. In Bhutan, for example, the Indo-Mongoloid peoples bear a strong cultural resemblance to the Monpa and Sherdukpen people living in neighboring Arunachal Pradesh. Their economies, architecture, and simple technologies are remarkably similar. All the eastern tribes share animistic traditions, some of them uniquely influenced by Tibetan Buddhism. They practice forms of nomadic horticulture, slash and burn the forests to clear small farm plots, and are known as swidden cultivators.

The earliest inhabitants living in the mountains near the eastern border of Nepal are known as the Lepchas, and are thought to also have come initially from Assam. Today, the Lepchas constitute an important tribal minority in Sikkim. They speak a Tibeto-Burmese language and, like some of their neighbors in Nepal and Bhutan, still practice elements of the ancient Bon religion. The Lepchas perhaps are known best by the fact that they live at the base of the spectacular Mount Kanchenzonga. At 8,598 meters elevation, it is the third highest peak in the world and one of the most beautiful mountains in the entire Himalayan range. For the Lepcha people, moreover, Mount Kanchenzonga in its entirety is sacred.

Much of the territory in eastern Nepal and Sikkim, meanwhile, especially that which borders on the valley of the Arun River, was called Kirat, and the region's indigenous Rai and Limbu tribes are popularly known therefore as the Kirati people. The Sanskrit literature actually identifies all early settlers of the eastern mountains as Kirati people, inferring a pan-Himalayan culture type for at least the eastern portion of the range. The lack of archaeological evidence in the eastern Himalaya, however, restricts our knowledge about the society or economy of the region's aboriginal people.

Oral narratives of the eastern range tell about a series of successive conquests from Burma and Assam. The scattered Kirati populations had organized themselves into small tribes who often fought one another for territorial control. The tribes were highly mobile, because the people hunted and gathered food from the forest. They needed large amounts of accessible wilderness to survive and waged battles with each other to secure territorial claims. Apart from a few local chieftains, who occasionally sought labor recruitments in order to build their forts, the eastern tribes mainly lived autonomous lives in small but self-sufficient bands.

The loose tribal arrangement in the eastern Himalaya, and the absence of strong regional rulers there, forestalled the development of a centralized system of control in the prehistorical period. In the western portions of the range, though, the indigenous populations often united under local leaders and fought wars over territory in broad regional alliances. Several periods of political consolidation occurred in the western mountains of India, culminating around 1400 B.C. when tribes from the Lahaul, Spiti, and Kinnaur districts united to fight in the great Bharata war.[6] Gradually, the mountain tribes were subjugated by the imperial expansions of lowland rulers. Monuments, rock inscriptions, and other historical records show that by the sixth century B.C., much of the western Himalaya was already under the control of empires that were centered in the Ganges plain. The greatest of these was the Gandhara civilization, which flourished under Ashoka's rule during the fourth century B.C. It penetrated into Kashmir, along the upper Indus River, and extended as far north as Ladakh and Zanzkar. The extension of the Indian empires into those areas facilitated a process of cultural and political interchange between the plains and the mountains that continues to this day.

Much of the Great Himalaya and the plateau zone, meanwhile, was settled from the north by the Tibetan bhotiya peoples, beginning perhaps as early as the start of the Christian era. The Tibetan migrations occurred sporadically in most places, and involved only small population movements. In Zanzkar and Ladakh, however, a full-scale invasion by Tibetans took place in the middle of the seventh century A.D. Rock inscriptions along the upper Indus River show that the Tibetan invaders met small tribes of bronze-age hunters living in the highland valleys. The hunters came most likely from the steppes of central Asia. The archaeological evidence and the ethnic makeup of the contemporary settlements in eastern Ladakh suggest that the arid northwestern Himalaya had been inhabited originally by Dardic peoples (who were of central Asian origin).[7]

The Dards in Ladakh were subdued by the Tibetans in the seventh century as part of Tibet's deliberate policy of territorial expansion. The Tibetan empire was at its zenith then, and in direct competition with the Chinese for control of much of central Asia. Under the rule of the great Buddhist king Songtsen Gampo and his successors, the Tibetans made it as far west as eastern Turkestan before they were stopped at the Oxus River by the allied armies of Arabia and China around the turn of the eighth century. The expansionist dreams of imperial Tibet never penetrated south of the Great Himalaya, and eventually the Tibetans retreated to the boundaries that existed prior to the sixth-century invasions.

The Tibetan dynasty in Ladakh was positioned against the Islamic forces in the west. It lasted for one thousand years. During much of that time, Ladakh was not an integrated part of the greater Tibetan state but acted instead as a dependency with colonial status. The Ladakhi court, however, annexed Lahaul and Spiti as additional minor vassal states. To escape riotous plunder, Spiti regularly paid tribute to its more powerful neighbor in the north. When this failed and the Ladakhis attacked, the people of the Spiti valley retreated to the high forts, called *dhankars*, that still cling to the cliffs above the Spiti River. The spectacular ruins at Dhankar Gompa and Khibbur, with their fortresslike bearings, reflect the valley's feudal history of plunder and subjugation. Meanwhile, the Spiti monastic center at Tabo, one of Buddhism's greatest shrines, displays magnificent fresco interiors, delicate courtyard stupas, and grand assembly halls attesting to the power and wealth of Tibetan monasticism over a thousand years ago.

The long reach of the Ladakhi court covered much of the arid western Himalaya until 1819, when the Indian king Ranjit Singh, based in Lahore, gained control of Kashmir and the legal right to extract tribute from Ladakh. The power of the royal court in Leh decayed as the imperial rule from the plains strengthened. Eventually, a Hindu rule was established in Ladakh which deposed the Tibetan aristocracy altogether and turned the entire northwestern region into a large vassal state of India. Despite the political changes, the cultures of Ladakh, Zanzkar, and Spiti maintain prominent Tibetan characteristics.

At the time when Tibetan forces were invading the western Himalaya, the armies of Songtsen Gampo took control of the central range in Nepal. By A.D.640, much of the country as far south as the Kathmandu Valley was under his rule. The Tibetan bhotiya groups now living in Nepal, as well as the descendant tribes of Tamang, Gurung, and Magar peoples, are believed to have originated in those widespread Tibetan invasions. Elsewhere, the arrival of the Tibetans was on a more modest scale. They settled sporadically with their livestock in the high valleys. The Dolpo and Mustang regions of Nepal and the border areas in Sikkim and Bhutan which adjoin the Chumbi Valley contain Tibetan valleys that have been populated since that earlier period. The assimilation of Nepal's remote regions into the sphere of the Kathmandu court by the end of the eighteenth century brought the distant inhabitants under nominal Nepalese control, but the landscapes and societies of Dolpo and Mustang still tie those places to the north, and their cultures are highly reminiscent of those found in Tibet.

A procession of monks at the Tabo Monastery, Spiti, India, marks the occasion of an important religious festival. The Tabo Monastery is over 1,000 years old and is a splendid example of Tibetan monastic architecture and religious artwork. Photo by D. Zurick

It is often speculated that the Thakali bhotiya, who live in the southern part of Mustang along the Thak Khola Gorge, came to the central region of Nepal from the Khasa-Malla kingdom centered farther west in Jumla. If so, they are culturally distinct from their immediate neighbors to the north. In northeastern Nepal, the Sherpa people trace their origins to the Tibetans who crossed over the Nangpa La into the Khumbu region sometime around A.D. 1533. The proto-Sherpa people apparently were fleeing their Kham homeland, located 1,300 miles away in the humid, forested regions of eastern Tibet. The Sherpa name, which they pronounce "Shar-wa," means "people of the east" in the Tibetan language. The Khumbu region is thought to have been an important extension of the Tibetan cultural sphere from early times, with flourishing trade and religious links to Lhasa. A short distance to the east of Khumbu, along the upper stretches of the Arun valley, are the settlements of other bhotiya groups who have ties not only with Tibet but also with the older, aboriginal cultures in Assam.[8]

Although the environmental impact made by the early settlers remains unclear, and will likely never be known, their use of fire for landscape management was definitely widespread, at least in the eastern regions. That possibly explains in part the contemporary pattern of forests in such places as Khumbu, in Nepal. Geographer Barbara

Brower has reported that the Sherpa still use controlled burning to manage some of their pasture land. Quite possibly, from the beginning of human settlement, rangeland management in Khumbu has included the use of fire.

The tribal populations of the eastern Himalaya have always relied heavily on hunting and foraging, but their subsistence activities are likely to have had little impact on the absolute numbers of game animals or on the productivity of the natural habitats. The shifting cultivators inhabited the lower mountain slopes and cut patches in the forests for their farm plots, but their low population densities prevented irreparable land degradation in most places. It is only in most recent decades, when population densities reached high levels and wildlands became restricted, that shifting cultivation has become an unsustainable form of agriculture in the mountain areas of modern-day Sikkim, Bhutan, and Arunachal Pradesh. Contemporaries concerned with land degradation in these and other Himalayan areas invariably name shifting cultivation as one of the main contributors to land degradation.

In the northwestern Himalaya, meanwhile, the historical circumstances surrounding the use of the land were quite different. Pressure to meet the tributary demands of the Tibetan rulers, as well as the subsistence demands of the local inhabitants, put a burden on the fragile plateau environments during even the earliest times. The Tibetan monarchy was initially established in Ladakh to capture the lucrative trans-Asiatic trade, but the royal courts were supported by local taxes on grain, butter, and forest products. As the power of the Ladakh court grew, it sent armies into the adjoining regions of Spiti and Zanzkar to collect tribute, which caused the valley inhabitants to farm the land more intensively and to exploit the medicinal, game, and timber products of the highland forests.

Dynastic Origins

The feudal Tibetan states were restricted in the central Himalaya to the arid northern plateau and to the protected high valleys. In those places, Tibetan societies functioned independently of Hindu domination from the south. Dolpo, Mustang, and Khumbu, for example, operated autonomously until as late as the mid-twentieth century. Their main allegiances and trading partnerships were with the Tibetans in the north, although some communities also brokered the caravan trade that existed between Tibet and India.

As early as the fifth century A.D., a group of powerful families connected to the nobility of India joined in a power block to control the mountainous area that today corresponds to much of central Nepal. Rock inscriptions, coinage, and other documentary evidence show the establishment of a Licchavi dynasty in A.D. 464 in the Kathmandu Valley. At its peak, Licchavi rule may have extended from the Kali Gandaki River in the west to the Kosi River in the east, thus covering much of the territory that is contained in the current outline of Nepal. But the Licchavi empire had no clearly demarcated boundaries, and the peripheral regions were poorly integrated with the rest of the kingdom, so the formal outlines of the empire are blurred. For the most part, Licchavi rule did not extend into the Tibetan areas.

The Licchavi dynasty borrowed much from Indian Sanskrit civilization and reached its zenith in Nepal in the seventh century, when its artisans blended the Hindu and Buddhist traditions in the Himalaya to create extraordinary artwork, temple architecture, and handicrafts. In order to maintain the artisan classes, as well as the large number of imperial functionaries, the Licchavi rulers promoted intensive farm production in the Kathmandu Valley and surrounding hills. The production was then taxed heavily by the state. Elaborate exchange systems and infrastructures were established during the prosperous period of Licchavi rule. Irrigation systems were built to boost agriculture, a road system was laid out, and the valley population was organized into a proper tributary society. The Licchavi rulers established political and military relationships with both Tibet and India, and Kathmandu became an entrepôt between the two countries. Thus these rulers laid the groundwork for geopolitical alliances that continue to describe Nepal's foreign policy. The immediate impacts of Licchavi rule were felt strongest in Kathmandu and along the major trade routes that led north and south from the valley. The control of trade, in fact, solidified the financial base of the empire.

Beyond the Kathmandu Valley, the mountains were inhabited by numerous ethnic groups who lived outside the control of the Licchavi authorities. The Limbus and the Rais in the east, the Magars and

POWER AND TERRITORIAL CONQUESTS

The Kathmandu Valley in Nepal.
Photo by P. P. Karan

Gurungs in the central mountains to the west of the valley, and various bhotiya groups inhabiting the high mountains maintained fairly autonomous lives. They occupied the territory that eventually would become consolidated under modern Nepal, and still constitute important tribal minorities in the kingdom.

The Licchavi dynasty formally ended in the late ninth century and was followed by several centuries of political instability and scattered feudal rule. In the eleventh century, the Malla kingdom emerged in the Kathmandu Valley as a contemporary to the Khasa-Malla empire centered farther west in Jumla. The Kathmandu Malla kingdom was supported by the rich paddy agriculture in the fertile valley, by Indo-Tibetan trade, and by the flourishing production of handicrafts. Intensive cultivation of the valley was also encouraged by the Mallas, to support the burgeoning urban population. Although most of the valley residents lived in small farming villages, the construction of temples, ashrams, government buildings, roads, and bridges that occurred during the Malla period gave to the Kathmandu Valley a look of high urban civilization.

Throughout the Malla period, several contesting powers were present in the middle-mountains areas and in the tarai regions. There is evidence of small-scale invasions from time to time into the Kathmandu Valley from the west, but none of them lasted for very long. The Muslims were in the Ganges plain to the south and by the fourteenth century they too invaded Kathmandu for a short time. Waves of Hindu migrants, including high-caste Brahmans fleeing the Muslim conquests, arrived in the Himalaya, adding to population growth in much of the region and bringing Hindu forms of religion and society to the mountains.

Some of the Hindu refugees who arrived in the western and central Himalaya came from Rajasthan and claimed royal descent. Unlike the earlier immigrants, who were from small-scale tribal systems and merely sought refuge, the Rajasthanis were noblemen and came with armies to subjugate the mountain people. They took the name Thakuri in western Nepal; elsewhere in the Himalayas they were known as the Rajputs. The Rajputs displaced the khasa ruling families in western Nepal and the other local rulers elsewhere in the mountains. They established numerous small kingdoms, which eventually extended from Kashmir to Sikkim, and formed new political alliances both with the existing tribal leaders and among themselves. The Rajputs organized the medieval landscape of the Himalaya into a gigantic mosaic of princely states whose chief ambitions were to gather wealth from the mountains and increase the power of the kings, known as the *rajas*.

The Pemayangtse Monastery is one of the oldest and largest Buddhist centers in Sikkim.
Photo by D. Zurick

The Medieval Era of Mountain Princes

5

IN THE FIFTEENTH AND SIXTEENTH centuries, twenty-two principalities, known as *rajyas*, became established in the western Himalaya within a loose alliance known as the Baisi. The rulers were all descendants of emigrants from Rajasthan. Their combined control extended from central Nepal to Garhwal. In the eastern Himalaya, an additional twenty-four Rajput kingdoms formed a second alliance known as the Chaubasi confederation. It reached into Sikkim. Other petty princes ruled elsewhere in the mountains, so, in total, eighty separate Hindu princely states controlled the mountains during the medieval period. The productivity of the land and the payment of tribute, the shifting alliances of the vassal tribes, and the patronage of the Muslim rulers in the southern plains all contributed to the fluctuating fortunes of these mountain princes.

Rajput political control of the Himalayan landscape created huge revenue demands on the villages and led to numerous territorial wars. These events, in turn, caused local people to lose control over their land and led to intensified use of mountain resources. The preoccupation of most rajas was with the acquisition of power and wealth, both tied directly to the productivity of the land, and hence they exploited to their fullest advantage the natural resources of their mountain kingdoms. The collection of revenues from peasant production bolstered the security of the petty kingdoms, since it provided the means for maintaining the princely armies. Moreover, to secure favorable trading arrangements in the region, the Himalayan principalities paid tribute to Muslim rulers who controlled the lowlands in the south. The plains-periphery relationship that has come to dominate regional affairs in the Himalaya may be properly attributed to this early formative period in the Himalayan political economy.

One of the most influential and wealthy of the small Himalayan kingdoms was in Salyan, located along the lower Bheri River in the western Mahabharat range, under the rule of the Phalabang raja. There, the villagers lived on household plots of land, among thatch and mud communities scattered across the hillsides. They paid allegiance to the raja, whose legendary wealth was tied to the tribute he collected from the rice harvests in the Sarda, Bheri, and Luwam river valleys. Irrigation was practiced by the villagers on the fertile river terraces, and the warm climate supported two crops a year. From the brick and wood fort that is situated still on a high ridge overlooking the village of Phalabang, the raja governed all of Salyan, and maintained powerful relations with the neighboring rulers in nearby Pyuthan and Dang, located east and south of Salyan, respectively.

A degree of security was enjoyed by the Himalayan farmers under the various rules of the independent rajas, so long as they continued with their tribute payments. Labor was occasionally requisitioned from the peasants in order to build temples, forts, bridges, and other royal infrastructures. Despite these demands, most peasant households had sufficient land and labor to meet their own food needs. Those favorable circumstances declined, though, with the unification of Nepal in 1769 under the Gorkha king Prithvinarayan Shah and the subsequent dissolution of the Baisi alliance. By 1789, most of the western Himalayan region was annexed under highly centralized

Gorkha rule, and new land tenure systems were introduced which directed the revenue away from the rural communities and the local rajas, toward the central government offices located in Kathmandu. As a result, the autonomy and influence of the country princes declined, although they continued to maintain their titular power, and the peasants could no longer depend upon the local rajas for protection against the abuses of the Kathmandu throne.

In the eastern Himalaya, meanwhile, the itinerant Tibetan herders from Kham, the displaced Tibetan noblemen, and the persecuted Buddhist lamas who fled from the power struggles in Tibet settled in the lush highlands located south and east of Mount Kanchenzonga. The Tibetan immigrants established monastic centers in Sikkim and in Bhutan, from which they dominated the indigenous Lepcha people.[1] In the seventeenth century, the competing Tibetan powers united in Bhutan under the state rule of a solitary spiritual king, called the dharma raja. In Sikkim, the aristocracies forged the Chogyal monarchy, which operated under the tutelage of the powerful Tibetan monasteries.

In both places, a land tax and corvée labor were levied on the peasantry, leading to such oppressive conditions that many people fled the high mountains to seek protection in the Indian lowlands and the southern Duars Plain. The dharma rajas also controlled trade with Tibet and the Assamese lowlands. Bhutan eventually managed to gain its independence from Tibet, so by the late medieval period it existed as an autonomous country.[2] Sikkim, meanwhile, remained bound to Tibet through military assistance, the intermarriage of royalty, and land deals. Wealth was gained for the Sikkim state by revenue collectors according to their land assessments. Much of the land payments, though, went directly to the monastic coffers in Lhasa or to other political centers in Tibet.

The religious traditions introduced to Sikkim by the bhotiya people integrated the animistic beliefs of the indigenous Lepcha people with the Buddhist doctrines of Tibet. This gave to Sikkim its unique spiritual order, which still includes the worship of numerous local deities who live in the forests, streams, and caves. The greatest land deity is Kanchenzonga, the mountain god revered by all Sikkimese. Near the mountain is the remote Yuksom Plateau, which is thought to be in the shape of a holy mandala and hence is some of the most sacred ground in Sikkim. The deities who define for the Sikkim people a supernatural world thus integrate on the deepest levels of human consciousness the lives of the indigenous inhabitants and the history of the land.

Much of the success of the Chogyal rulers in Sikkim rested on their ability to unite diverse ideas about native culture, sacred territory, and national identity. This union was breached, though, with the arrival of the Nepalese settlers; the imperial claims of the British, who were in India as colonialists; and the influx of Tibetan *rimpoches* (high-ranking Buddhist clergy) who established the great monasteries, such as Pemayangtse and Rumtek, which still are prominent in the daily political affairs of Sikkim. The Tibetan lamas gradually usurped the authority of the traditional Sikkimese spiritual leaders, called the Gomchens, who had mediated village life from their mountaintop retreats. Antagonism came quickly to dominate early relations between the indigenous Lepcha people and successive waves of newcomers, resulting in a gradual loss of native land rights during the political formation of modern-day Sikkim. Only recently, with their sacred land threatened by proposed hydroelectric schemes along the Tistha River, have indigenous groups begun to assert themselves in the national development debate.

Farther along the northeastern mountain frontier, among the ridges above the Siang and Subansiri rivers of Arunachal Pradesh, the isolated tribes continued to live outside the pale of the Himalayan empires. The wild mountains of the Paktai region, where some twenty-five distinct tribal groups still live, were never subdued by Rajput princes or Tibetan kings. The local Ahom kingdom, based in the Assam Valley had nominally influenced the lives of the mountain tribes, but the isolation of that region throughout the Middle Ages continues into the modern era. It remains as one of Asia's last outposts.

The Gorkha Empire

One of the most powerful of the Hindu medieval monarchies to emerge in the central Himalaya was the Shah kingdom of Gorkha. In 1769 the Gorkha rulers had conquered the Kathmandu Valley and had begun a period of imperial expansion that was eventually to extend their Himalayan rule across an area of almost 1,500 kilometers, stretching from Garhwal to Sikkim. The strategy of the Gorkhas involved controlling the important trade routes between Tibet and India. That was most significant because China at the time was asserting its suzerainty over Tibet, and the British were in ascending control of much of India. Hence, the Gorkhas felt that combined military and commercial unification of the central Himalaya was necessary in order to forestall the territorial ambitions of the Chinese and the British.

The geographical expansion of Gorkha rule rested on how quickly the various Rajput principalities could be subdued. Some of them acquiesced quietly, negotiating favorable terms of trade and revenue collection in exchange for giving up their sovereignty to the Shah kingdom; others of the princely alliances resisted the Gorkha intrusions. The latter cases met with overwhelming defeat, since the central armies of the Gorkhas were much superior to any of the forces that could be mustered by the petty independent Hindu kings. By 1789, much of the western region, all the way to Kangra and the Sutlej River, was annexed by the Gorkhali empire. Along with most of Nepal, the western mountains thus came under the direct administration of the Kathmandu court, known as the Durbar.

The rajas who had fought unsuccessfully against the Gorkhas were either killed outright or fled to the plains to seek protection under the British colonialists. Those who chose to remain under the authority of the Gorkhas retained some autonomy over the internal affairs of their kingdoms. Indeed, the titular role of the petty kings continued in Nepal until as late as 1961. But the power they wielded over their subjects and territories quickly diminished as the administrative arm of the Kathmandu Durbar expanded into distant conquered territories. The vital matters of land and revenue collection were among the first orders of business for the Shah emperor.

Gorkha policies introduced a rigid definition of land tenure throughout the central Himalaya and initiated widespread land transfers between villagers and the state. The Gorkha rulers issued royal decrees throughout the kingdom, conceding large amounts of land to their military officers. In the western range, small cultivators discovered that their unregistered lands would be confiscated and that they could continue to till their soil only as tenant farmers under the new *adhiya* system of land tenure. The adhiya tenure appropriated vast amounts of land for the benefit of certain favored classes, notably the

Brahman, Chetri, and Thakuri communities. A 50-percent tax on farm production was assessed, although landlords often demanded a good deal more. Fixed rent payments were introduced under a new taxation system in 1812, adding to the burden of the peasantry by removing guarantees against poor harvests. This led to increasing indebtedness among the peasantry, exacerbated by the high interest rates charged by the local moneylenders.[3] Lending rates as high as 50 percent are still found among the Thakuri and Chetri moneylenders in the villages of western Nepal and in the adjoining areas of Kumaun and Garhwal in India.

The impact of postunification politics on the land and the environment extended beyond tenure relations and entered directly into land management. Cultivators who could not pay the high rents under the new systems of tenure were evicted and replaced by outsiders who could. Thus, the customary village land rights were violated. In some cases, the land that was relinquished was not taken over by outsiders, but reverted instead to wasteland. A common and potentially more insidious event, though, was the settlement and intensive cultivation of all open lands. That effort was encouraged by the Gorkhas in order to increase revenue production for the state.

The period from 1790 to 1810 represents an important early phase of land clearing in the parts of central and western Himalaya that were under Gorkha rule. It occurred as a direct result of the adverse tenure relations and the exercise of external political will, and not of local population pressure. To a large extent, the current land problems across much of the Himalaya date back to this period. Intensive land clearing and resource extraction, as well as inequitable land ownership, resulted from Gorkha policies, showing a clear political origin for some of the environmental problems that trouble the region today.

The need for revenue to maintain the large standing Gorkha army provided the major impetus for the land-clearing efforts. The construction of terraces and supporting structures for agriculture, such as irrigation canals, required the use of forced labor conscripted through the *jhara* (military) system. Compulsory labor and semi-enslavement arose mainly because the peasants had sufficient access to land for their own needs and so felt little compulsion to engage in additional land clearing unless forced to do so. With the imposition of state control, local land needs were often ignored. The intensified use of mountain lands not only increased the level of farm production but also further promoted the extension of state control over the local inhabitants. The Gorkha administration, through its local tax officers, military garrisons, and petty bureaucrats, kept a close rein on the people and on the resources located within their domain. This dominion extended to the way villagers farmed the land and managed the forests. The early loss of forest areas in central and western Himalaya can be attributed in part to the extractive policies of the Gorkha rulers, who saw in the domestication of the wildlands the potential for state profit,[4] and so the Himalayan forests felt early on pressures for commercial development.

Agricultural- and forest-land allotments were awarded directly to military officers for service to the Gorkha empire. The rulers allowed the military estates to collect revenue from the tenant farmers and to demand labor from the villagers for the purposes of building and maintaining roads, forts, and bridges. The military estates also recruited from Kathmandu skilled workers and artisans, who settled near the garrisons and lived as professional persons. Such settlements gradually formed provincial centers, occupied also by Newari businessmen from Kathmandu, and eventually became important regional market towns.[5]

The conversion of the forests to cultivated land was accelerated when direct incentives for land clearing were offered to the peasants by the Gorkha regime. Tax breaks and contractual land rights were given to farmers who cleared forests for agriculture. Under those incentives, land ownership reverted to the cultivator after an initial three-year rental period. The allure of such a strategy for the peasants, as indicated by their widespread participation, was nothing more (or less) than reestablishing the indigenous rights over land that they had enjoyed prior to the new land policies of the Gorkhas. But the result was a large-scale clearing of landscapes all across the mountains.

The land-settlement process and the expansion of agricultural production led eventually to the need for more exact accounting of revenue on contract lands and to the new jhara system, which recruited village headmen to collect the revenues in a more systematic fashion. These changes enhanced local control over village affairs and introduced a system of tribute payment that more closely resembled the older, traditional relationship between peasant and chieftain, rather than the more recent one between subject and state. But the net economic and environmental change was minimal; resource wealth still was siphoned off by the central government and the clearing of land continued to increase. Moreover, as the authority of village elites over local finances—and hence over peasant production—expanded, so too did power accrue to those persons. By virtue of their positions as revenue collectors and moneylenders, the village elite became even more entrenched, and social stratification developed in the rural communities.

Individual land rights, meanwhile, were largely reinstated, though not necessarily benefitting the same households, and were periodically enforced by royal decree. By the middle of the nineteenth century, a customary law provided households and individual farmers with more control over land and cultivating rights. By the third quarter of the nineteenth century, the peasants once again had ownership rights and could buy, sell, and inherit land. When land rights were secured, households invested in family-based land clearing. Populations slowly increased in the central Himalaya during the late nineteenth and early twentieth centuries, leading to even more intensive farmland expansion during that period.

The pattern of land occupancy in Salyan shows a close correlation to the broader trends described for the central and western Himalayan regions. Prior to the Rajput infiltrations, the Phalabang forests were used by bhotiya herders from the north and by foragers, hunters, and shifting cultivators who came from the southern and eastern directions. The extent of local settlement during this early period is difficult to determine, but the bands of hunter-gatherers, called *raute*, that once lived in the region are thought to be part of a much larger group of people who had settled thoughout the Bheri and Karnali valleys. One of the settlements near Phalabang is called Rautechaur, indicating the early presence of the raute group in that place. Under the earliest foraging and cultivating systems, the land clearing was minimal. Settlements were greatly scattered around Phalabang, and people ranged widely among the forests in their pursuit of game, fish, and wild plants.

The arrival of Hindu settlers in the seventeenth century brought livestock and grain farming to the valleys of the Sarda and Luwam rivers. Rice fields expanded along the terraced slopes of the lower river valleys during the reign of the Salyan rajas. By the middle of the eighteenth century, Phalabang was a well-established village ("bang" referring to a farming place on the side of a hill). When the Salyan area was formally declared a principality in 1766, it contained some ten thousand houses. The strategic site of the Phalabang fort commands magnificent views of the Sarda and Luwam river valleys, the surrounding ridges, and the main trails entering and leaving the territory. Beginning in 1842, when residency by the rajas was first established in Phalabang, the raja oversaw his political realm and collected revenue from that mountaintop retreat.

The expansion of the Gorkha empire in the eastern part of the Himalaya took a somewhat different turn. The Shah kings did not at first subjugate the Kirati tribes as they did the Hindu peoples in the central and western parts of their empire. Instead, the Kirati chieftains retained political control over their tribes by entering into treaty agreements with the Gorkhas. The treaties nominally recognized Gorkha suzerainty, but allowed the Kirati tribes in the eastern range to retain their traditional system of communal land ownership, known as *kipat*, which was vitally important for maintaining the systems of tribal culture. In Sikkim, which the Gorkhas invaded in 1788, the ruling abbot, Tenzing Namgyal, was forced to pay tribute to the Kathmandu rulers as a vassal state. Sikkim maintained that status until 1815 when the British defeated the Gorkhas. Sikkim then became an independent state, albeit under British influence, until 1861, when it was made into a direct British protectorate. The rulers of Sikkim had come initially from the Chumbi Valley, which is located across the border in Tibet, and paid their allegiance to the Tibetan aristocracy in Lhasa. Under the British protectorate, though, Sikkim leaders were forced to live in Gangtok, where they could be more easily managed by the British crown. A British political agent was stationed in Gangtok as a reminder of the imperial power of Great Britain even in that remote place.

The Gorkha policies were meant to pacify the natives and allow the migration of Hindu populations into the lower mountains. The high-caste Hindus who settled in the eastern region gradually gained land rights in the form of debt repayments or rent. As a result, kipat land increasingly came under the control of Hindu rural elites. The range of territory available to the Kirati tribes diminished, and some quit the extensive practices of hunting, shifting cultivation, and foraging, and became sedentary farmers.[6] When the Rana family in Kathmandu took control of the empire from the Gorkha rajas in 1846, it demanded cash tributes from its eastern subjects rather than in-kind payments. Consequently, tribal cultivators were forced to borrow money from the Hindu moneylenders or to mortgage their lands to pay the taxes. That increased their indebtedness. The autonomy and well-being of the indigenous tribes in the east deteriorated as the power of the Hindu elites in the countryside increased. A general impoverishment of the eastern mountain tribes occurred which sent many people fleeing into Darjeeling, Sikkim, and Bhutan.

Thus, along with the cultural erosion of the Sikkim and Bhutan tribes, there occurred an actual physical displacement of people. Tribesmen from neighboring Nepal moved into the region in search of land and new opportunities. By the latter part of the nineteenth century, over half the population in Darjeeling was of eastern Nepalese origin. The flow of Nepalese migrants into Sikkim was also great, so that by the end of the nineteenth century the Sikkimese authorities had to take strong measures to staunch it. These measures led to ethnic separatist feelings and to human-rights abuses. The Sikkim population today is predominantly of Nepalese ethnicity, ruled by the Chogyals of Tibetan descent. The indigenous Lepchas, meanwhile, have become minorities in their own country.

Nepalese also settled in Bhutan, where they quickly became a sizable group. Bhutan recently adopted new land-ownership rules that have pushed people of Nepalese origin off their lands. That policy resulted in the early 1990s in large-scale refugee migrations back to eastern Nepal, where people now live in tent encampments or with relatives among the rural villages of Ilam. The eastward spread of Nepalese migrants, had, however, eventually extended all the way into Assam, where by 1921 they numbered over seventy thousand inhabitants. The decisions made by Nepalese to move to the eastern ranges were not lightly made, since they uprooted people's lives and properties; they illustrate the serious levels of cultural and territorial conflicts that prevailed in eastern Nepal between the Hindu elites and the native residents during the middle and late nineteenth century.

Such land disputes were not restricted to the eastern mountains of Nepal. Mountain tribes residing elsewhere in the Himalaya endured similar threats to their land and territorial rights.[7] Land alienation was most common where the Hindu groups settled in large numbers. Bhutan closed its borders in part to protect the interests of its citizens. History and geography, meanwhile, combined in the far eastern frontier to protect the hill tribes in Arunachal Pradesh from the encroachments of the flatlanders. These tribes were periodically subjected to the abuses of the Ahom rulers in the Assam Valley, but mainly they were able to keep to themselves. Elsewhere, however, the extension of political control was commonly followed by the loss of indigenous land and resource rights.

The southward expansion of the Gorkha empire onto the Himalayan piedmont plain during the early eighteenth century was accomplished primarily through treaty alliances with the existing lowland elites. That strategy, however, was not entirely effective. Most of the tarai authorities were of Indian origin and distrusted the Gorkhas. The indigenous people of the lowlands, including the Tharu tribes of Nepal, were forest people and stayed away from both groups as best as they could manage. In Nepal, the Maithili Brahmans of the tarai also kept outside the sphere of direct Gorkha annexation. But the lowlands were of major economic importance to the Gorkha empire, and so its army continued to push southward onto the tarai until the Gorkhas unified much of it within their imperial boundaries.

The British colonialists based in Calcutta saw the Gorkhali invasions of the tarai as a direct threat to their own sovereignty along the border regions. Britain's interests also included the trans-Himalayan trade, which was then controlled by the Gorkhas. The British therefore waged a territorial war against the Gorkhas and in 1815 defeated them. As a result, the Segauli Treaty was signed; it gave the British the western Himalayan regions of Garhwal and Kumaun and large portions of the tarai. The eastern section of the Gorkha empire was restored to Sikkim, and the Gorkhas were left with the territory that roughly corresponds to present-day Nepal. The internal affairs of Nepal quickly deteriorated, though, under the later Gorkha rulers, and conditions of military chaos and royal scandal descended upon the country. In 1846, the control of a debilitated Nepal was seized by the aristocratic Rana family living in Kathmandu.

Throughout the medieval period, relations between nature and society in the Himalaya were shaped by state demands on farm production and on trade. The Ahom kingdom in lowland Assam, the princely states of Nepal, and the Tibetan kingdoms in Ladakh and the neighboring borderlands are all examples of extractive societies, but others existed as well. As a result of regional alliances and mercantile efforts, the Himalayan societies became well connected both to Tibet and the Indian plains. The kings and chieftains gained considerable wealth from the trade and revenue activities that took place in their domains. In addition to farm products, they exacted grazing taxes and royalties on timber and other forest products. In sum, the imperial rulers had ready access to the resource wealth of the Himalaya from an early date.

Initial state interventions in the resource systems of the mountains were facilitated by the slow formation of regional economies. The new political and economic systems, especially those of colonial origin, aimed to develop resources such as forests, water, minerals, and farm soil for commercial profit, and therefore have contributed greatly to the uneven history of environmental change in the Himalaya. The contemporary land degradation in the mountains is compelled in part by social arrangements that date to ancient times. The expansion of colonial empires during the nineteenth century brought to the Himalayan region additional European forms of land ownership, political control, and commerce. Such colonial patterns extended the possibilities for cultural and environmental disruptions as they brought to the region the insatiable demands of the world economy. The consequences for the Himalayan communities proved to be formidable indeed.

The Colonial Great Game

From the middle of the nineteenth century until Indian independence in 1947, in what has in Asia come to be called the Great Game, large sections of the Himalaya were disputed and fought over by the Russian, Chinese, and British powers, who sought to appropriate the mountains for their own respective colonial gain. The interests of the British in the region were centered upon the security of India's northern frontier and upon the possibilities for commercial development there. The Chinese and Russians, meanwhile, were advancing southward in their broad movements of empire building. These imperial efforts added new political subterfuges to a region that had a long history of power and territorial struggles. Like those of their medieval predecessors, the efforts of the colonialists affected not only the political and economic affairs of the mountains, but their environmental history as well.

6

British interventions began in the early 1800s as a series of explorations meant to formalize the Himalayan border of *pax Britannica* in India. The northern limits of the British colony in southern Asia were as yet poorly formed, and the colonialists were especially anxious about the possibilities of border aggressions from the north. British administrators sought fixed boundary lines for their maps. By establishing recognized authority in the mountains, they hoped to gain as well control over the lucrative trans-Himalayan trade routes. That was their primary commercial interest in the range. A secondary interest focused on exploiting mountain resources, especially mineral and timber wealth, and on developing the commercial possibilities of mountain agriculture.

Mapping the Imperial Frontiers

With the defeat of the Gorkhas in 1815, the sovereign territory of Nepal was fixed more or less as it exists today. A British official was allowed entry into Kathmandu following the signing of the Treaty of Segauli, but he was without authority in the Kathmandu court, or Durbar, and lacked even an advisory role. Nepal kept its full independence from Britain. When the Rana prime ministers gained control of Nepal in 1846, however, they saw that their long-term interests could be served by establishing favorable, if secretive, relations with the British. The recruitment of Nepalese Gurkha soldiers by the British regimental army was a key step in that regard. The Gurkhas solidified the bonds between the Rana elites and the colonial rulers. They became a form of diplomatic currency in the Ranas' efforts to safeguard their suzerainty over Nepal.[1]

During most of the Rana autocracy, though, Nepal closed itself to the rest of the world. The Ranas fraternized with the British administrators living in India, for whom they felt an aristocratic affinity, but kept all others at bay. They hunted tigers with English officers in the jungles of the tarai, and made frequent trips to London for pleasure and business. They came to depend upon the colonial offices in India in matters of trade, patronage, and succour. The concessions that the British made to Nepal's independence, formalized in the 1923 Treaty of Friendship, were based upon the recognition that Nepal provided for the British a convenient buffer zone along India's northern frontier. The problem of the international boundary with China was thus passed along to the Ranas.

In the far eastern and western portions of the range, disputed frontier lines crossed treacherous lands that were alternately claimed by the British, the Russians, or the Chinese. The demarcation of the actual boundaries in those frontier areas did not exist on paper maps, but they occupied a place in the mental maps of the colonial mind. Extreme deserts, rugged passes, and, in the case of the far east, dense jungles all kept the mapmakers away. Little regard was paid in any case to the native residents, who already had cultural claims to the land and a good idea of what it contained. The early British expeditions into the Himalayan frontiers, sponsored in part by the Royal Geographical Society, were meant to officially map the land for the first time and to describe the peoples who lived there.

The Uncharted Land

The British had gained nominal control over northeastern Himalaya in 1826 as a result of the Treaty of Yandaboo. That ended a short-lived Burmese invasion of Assam and solidified colonial claims to the mountains overlooking the valley of the upper Brahmaputra River. But the polyglot of insurgent jungle tribes, including the Sarchop and Chakmas, effectively curtailed widespread British authority in the area. The native people practiced shifting cultivation in these remote lands, and were suspicious of and hostile toward all outsiders, including the British; they fought to stay independent of the political will emanating from the plains.

Many of the smaller tribes paid minor tribute to the powerful Adi tribe in return for the use of land over which the Adis had gained customary rights in ancient times of warfare. When the European tea planters began to establish plantations in the Darjeeling and Mishmi hills in the 1800s, land conflicts between native people and settlers intensified, often requiring the intervention of British military forces.[2] The British, in their turn, attempted to stop the payment of tribute to the Adi tribe, thus increasing the overall hostility of the hill people toward the colonialists. That made expeditions into the remote regions all the more hazardous for military surveyors.

British garrisons were established in the eastern range in the latter half of the nineteenth century in order to protect the tea planters living there. By 1842, much of the upper Brahmaputra valley was administered, at least on paper, as an outlying district of Bengal. Colonial mapmakers were sent into the mountains in the 1870s to chart the frontier, but they were often killed or repelled by the tribes living there. The British eventually abandoned their plans for a full survey of northeastern Himalaya. Instead, they set up a series of remote frontier posts and established the Inner Line Regulation, which restricted travel by lowlanders into the interior mountains. The Inner Line established the pretense of an international boundary without the actual existence of one. The land that fell between the Inner Line and Chinese territory to the north became a no-man's-land for the Europeans, until the death of a British political officer in 1911.

THE MURDER OF CAPTAIN WILLIAMSON

The absence of British control over the Himalaya north of the Inner Line left the Assam region open to the devices of the tribal groups who lived there. The widespread intertribal warfare in the area worried the British, who viewed the local conflicts as a direct resistance to colonial control. To impress upon the natives their designs for the region, the British sent small reconnaissance expeditions into the mountains under the leadership of political officers such Captain Noel Williamson. Captain Williamson thought himself an ally of the tribesmen and regularly volunteered for the frontier journeys into the Mishmi hills and beyond, to make contact with the tribes and to map the rugged landscape there.

As the British struggled to gain control in the region south of the Great Himalaya, the Imperial Court of Peking was busy north of the mountains securing China's claim over Tibet. The Chinese were concerned that, despite the St. Petersburg Convention of 1907, which prohibited foreign intervention in Tibet without Peking's approval, both Great Britain and Russia had imperial designs on the vast plateau region. The British, on their part, were just as wary of China's being poised over the frontier of British-controlled Assam. They thought that the warm and fertile mountains south of Tibet must look tempting indeed to the Chinese. Moreover, the British had no idea how the hill tribes estimated the Chinese, but they knew that the tribes did not look favorably upon their own military installations.

And so it was under tense conditions when in the spring of 1911 Captain Williamson left his garrison for the forested hills north of the Brahmaputra River. After cutting his way for several days through the dense jungles of the hills, stopping en route at the tribal villages, Captain Williamson reached the village of Khomsing. He was murdered there on March 31. On the surface, the murder seemed to be simply another attack by a disgruntled tribesman against a colonial interloper. But further inquiries by the British showed that it had deeper overtones. Some Chinese agents had apparently infiltrated as far south as the village just to the north of Khomsing, and they were suspected of having ordered the attack on Williamson in order to further provoke tensions along the frontier. If that indeed was the case, their strategy was successful.

Following the attack on Williamson, the British sent well-armed military expeditions into the region to locate the perpetrators. They never were successful in identifying the killers, but the British were provoked highly enough by the incident to send new survey teams into the region in 1912 and 1913 with the goal of finally charting the mountain territory once and for all on British maps. The completed survey reports were then sent on to Shimla for review by Sir Henry McMahon. Basing his conclusions upon those reports, and on the recommendations of the Shimla Treaty Conference, McMahon fixed the frontier boundary in Arunachal Pradesh along the crest of the mighty Himalayan peaks between Burma and Bhutan. The eastern boundary, known as the McMahon Line, thus demarcated the northeastern frontier of the British empire.[3] Westward, however, the frontier boundaries remained a matter of contention.

Boundaries of the Dharma Kingdoms

When the Tibetans first conquered western Bhutan in the ninth century, the kingdom already had a loose cultural unity fostered in part by the ancient rule of the Assamese kings and in part by the missionary efforts of the religious teacher Padma Sambhava, who reportedly resided in Bhutan for some time during his extensive Himalayan missionary travels to spread Buddhism. The Tibetan influence over western Bhutan was maintained for several centuries by the monastic institutions and by the ruling families' close ties to the Tibetan nobility.

The easternmost part of Bhutan, though, remained loyal to the Assamese tribal chiefs.

In the middle of the seventeenth century, Lama Ngawang Namgyal consolidated political authority in Bhutan under the Drukpa sect of Kagyupa, a branch of Mahayana Buddhism. Drukpa was established as the state religion of Bhutan, and Ngawang Namgyal became the country's spiritual and temporal ruler. He became known as the Dharma Raja, and the country was named Drug Yul. Its unified territory was organized into several administrative regions, each with a local governor known as the Penlop and a chief lama. This territorial structure consolidated Bhutan as a national entity.

Internal rivalries among the monastic schools, a series of invasions by Tibet between 1644 and 1714, and its own aggression against Sikkim in 1700, shook the geographical foundations of Bhutan during the eighteenth century. Tibet negotiated a treaty in Thimphu in 1732 which gave to the Lhasa government some important tributary rights over Bhutan. The country nonetheless managed its own affairs as a sovereign country. Following advances into Sikkimese territory, Bhutan in 1771 sent troops south across the Bengal frontier. There they met British troops, who drove the Bhutanese back to their own territory. Disputes along the border continued until 1841, when the British finally annexed part of the Duars Plain in Assam and positioned the northern border of their empire along the approximate line of the present-day boundary between India and Bhutan.[4] Rival political factions in Bhutan fought over territory and prevented full unification of the kingdom until 1907, when Ugyen Wangchuk, the governor of the Tongsa district, assumed power.

Immediately to the east of Bhutan is the old kingdom of Sikkim. Prior to 1642, when Phuntshog Namgyal of Kham was consecrated as the first religious ruler of Sikkim under the title of Chogyal, the land was under the tribal governance of indigenous Lepcha inhabitants and immigrant bhotiya people. The coronation of Phuntshog Namgyal extended the country's territory to the Chumbi Valley in the north, to the Ha Valley in the east, to Ilam in the west, and to the Darjeeling plains in the south. The Kham-based Namgyal dynasty lasted, at least nominally, until 1974, when Sikkim was formally annexed by India. In the early eighteenth century, Bhutan conquered the land east of the Tistha River. Nepal gained Ilam at the close of the Gorkhali empire. In 1839, following the kidnapping and murder of several British officers, Sikkim was forced to cede Darjeeling to the East India Company, in return for which the British provided the Sikkim maharaja with an annual subsidy. To safeguard its trade and commercial interests, the government of British India in 1861 forced the princely state of Sikkim to accept British authority over both its internal and its external affairs. In the 1880s, Sikkim lost the Chumbi Valley during negotiations between the British and the Chinese authorities in Tibet. The border between Sikkim and Tibet, defined according to the geography of the upper Tistha watershed, was finalized under the protocols of the Anglo-Chinese Convention of 1890. The territory of Sikkim has remained essentially unchanged since then, although its control had passed from the Chogyals to the British, back to the Sikkim monarchy in 1947, and then to India in 1975.

The Great Game in the Indus Mountains

In his famous book *Kim*, Rudyard Kipling sketched the outlandish character of Mahbub Ali, a known horsetrader who was also registered by the India Survey Department as the mysterious agent "C.25.1B." The fictitious adventures of Mahbub Ali and his fellow spies under the service of the British in the northwestern Himalayan frontier closely resembled actual colonial exploits in the mountains during the nineteenth century. It was a period of high adventure, when the British, the Russians, the Chinese, and the northwest tribes all sought to assert control over the rugged territories of present-day Pakistan and India.

Little was known in the Western world about the frontier regions of the northwestern Himalaya prior to 1800. The early crossings of the mountains by Chinese monks in the seventh century were unfamiliar to the Europeans. Marco Polo had mainly circumvented the mountains during his famous mercantile journey in A.D. 1274, devoting only a few pages in his diaries to describing them. Even those passages are questionable, since some scholars now contend that Polo never went to China at all. Some Jesuit missionaries had visited the Himalaya in the 1700s, but they were concerned mainly with documenting the exotic religious practices of resident Buddhists. The European cartographers relied still on the fanciful maps of Ptolemy, which had the western Himalaya portrayed as a long line of skinny peaks, and those of

The summer headquarters of the British Raj was located in this building in Shimla. It now contains the offices of the Indian Institute for Advanced Studies. Photo by D. Zurick

Herodotus, who had populated the mountains with unicorns and gold-digging ants the size of foxes. Not long after the turn of the nineteenth century, three great empires were advancing toward the arid mountains of Ladakh, Zanzkar, and northern Pakistan. The Russians came from the northwest, through the central Asian kingdoms, by skirting the Tien Shan and Pamir ranges. The Chinese pushed across Tibet from the northeast. And the British were scouting north from India, up the tributaries of the Sutlej and Indus rivers. There was a great need for good maps showing the terrain and possible travel routes through the mountains. The three converging powers each sought to claim a foothold in the western Himalaya by virtue of being the first to reach there. This sparked a period of espionage and adventure that lasted for a century. Colonial thrusts and counterthrusts, steeped in the secretive journeys, the negotiated treaties, and the subterfuges of the frontier forces, made up the territorial exploits that came to be known as the colonial Great Game in southern Asia.

Not all the early Himalayan travelers, though, were in the service of the colonial empires. Some were simply opportunistic adventurers in search of forgotten lands and possible riches. Such was the case at least with a handful of European explorers. Those working for the imperial powers, though, were numerous, and they traveled through the mountains in the guise of natives, Arab traders, lamas, or Hindu pundits. They persevered under the tremendous difficulties imposed by the harsh terrain and climate, and often incurred the wrath of local rulers when they were discovered in their elaborate masquerades. Nonetheless, the frontier trophy of the western Himalaya was prized highly enough to compel a long series of daring explorations.[5] Beginning in Kashmir, which the British considered to be the jewel in the imperial crown, frontier expeditions eventually crossed the Zanzkar and Ladakh ranges, negotiated the valleys of the upper Indus River, and succeeded in breaching the Karakoram and then the Pamir mountains.

The British explorer William Moorcroft first crossed the Sutlej River in March 1820 en route to central Asia, ostensibly to buy horses for the East India Company. His journeys took him up the Beas River in Kulu Valley, across the Zanzkar and Ladakh ranges to Leh, and then on to Kashmir. It was evident, however, from his entourage, which included two Indian pundits measuring the entire route by counting footpaces on rosary beads, that Moorcroft's travels were more than a mere trading expedition. He spent ten months reconnoitering the Vale of Kashmir, but viewed his stay there only as a minor stopover in his more ambitious plans to reach Turkestan. His followers, though, had other designs on the valley. The British artist and spy Godfrey Thomas Vigne, who visited Kashmir in 1835, described its beauty thus: "All was soft and verdant even up to the snow on the mountain tops; and I gazed in surprise, excited by the vast extent and admirably defined limits of the valley and the perfect proportions of height to distance by which its scenery is characterized, rendering it not unworthy of the rhyming epithets: Kashmir bi nuzir (Kashmir is without equal), Kashmir junat puzir (Kashmir is equal to Paradise)."[6]

The glowing accounts of Kashmir recorded by early visitors such as Vigne led to its becoming the geographical centerpiece of British ambitions in the western Himalaya. The explorers not only described the natural beauty of the valley but also explained its critical role in the caravan routes that led to western Tibet. They pointed out how Kashmir and the neighboring district of Ladakh held the transit keys to the vast marketplace of central Asia. They intimated the vulnerability of the British frontier by describing the passes through the mountains that would enable invaders to cross into colonial territory from the north. Kashmir could become India's stronghold in the north. That was a concern not lost on colonial strategists as far away as London.

Meanwhile, Russian advances southward into the mountains during the 1700s and 1800s had led to the acquisition of more than 7 million square miles of territory in the Pamir and Tien Shan mountains. The movement of Russian explorers and army troops onto the northern slopes of the Himalaya provoked great concern among the British leaders. As a result of the Segauli Treaty and the retreat of the Gorkha rulers into Nepal, the British had acquired a vast tract of land east of the Sutlej River in Garhwal and Kumaun. Now it was time to shore up imperial boundaries in the extreme northwestern part of the range.

The control of the territorial frontier was considered to be vital to the security of the entire British raj. As explained by Lord Curzon of the Royal Geographical Society, frontiers "are the razor's edge on which hang suspended the issue of war or peace and the life of nations."[7] Establishing such control in the Himalaya, however, was to remain an elusive goal for quite some time. Compounding the difficulties of the terrain and the warrior tribes—including the Pathans and the Sikhs, who were only nominally subjugated when the British won the Sikh Wars of 1845 and 1848—was the lack of cartographic accuracy about the region. For example, no one knew exactly where the boundaries between India and Tibet lay. They appeared on no maps. Nor were the intervening topography or mountain passes well mapped. The movement of troops and the extension of British authority over the northwestern trading routes were stayed by the cartographic omissions. That was to change, however, beginning in 1850, when the Great Trigonometrical Survey of India reconnoitered the entire Himalayan range.

The surveyors of the Himalaya were both accomplished geographers and daring adventurers. Their task was to situate the mountains within the great mathematical spheroid of India, first calculated by Sir George Everest in the early 1800s, and to subdue the mountains by their maps. The survey groups, armed with theodolites and plane tables, climbed mountains that had defeated the most accomplished alpinists, and crossed terrain that had been previously untraversed by Europeans. They set heliotropes on the tops of high summits so that sightings could be made by surveyors from down below. In the dangerous tribal territories, Indian assistants, disguised as wandering ascetics, measured the long distances with prayer wheels and wooden rosary beads. By the end of the nineteenth century, all the major western Himalayan peaks were triangulated, the rugged terrain was mapped, and the frontier lines were drawn to show the northern boundary of the British Indian empire. It was a remarkable accomplishment for the colonial cause.

The explorations and maps were crucial to the British for a number of reasons. They gave an order to the imperial space that was based upon European principles of cartography and geometry, which allowed the British to construct a rational image of its Indian empire.[8] By defining the Himalayan periphery, the maps solidified the power of the imperial core. They formally extended British hegemony to the princely states and to the mountain tribes, showing them to be components of a unified empire. The maps measured the trading routes that were such a lucrative prize in the Himalaya. They allowed a more precise reckoning of land productivity and became the basis for assessing revenues and taxes; finally, they showed the wealth of mountain resources, which was eventually to be exploited by commercial entrepreneurs. In sum, the frontier maps were the colonial keys to subjugating both the lands and the peoples of the mountains.

Colonial Mercantile Interests

Two great trading routes wrapped around the Himalaya in a vast commercial embrace. The famous Silk Route into central Asia, connecting China and the Middle East, crossed the mountains in the western sector. A second one connected India and China in the northeast, passing through Yunnan and the upper Brahmaputra valley. The latter route traversed India via the Grand Trunk Road and tied into the Silk Route near the Khyber Pass. Between those great trading avenues, numerous caravan circuits crossed the Great Himalaya on high mountain passes and followed along the river valleys. The possibility of colonial control over those lucrative passages required both the presence of military garrisons and the establishment of relations with the local rulers.

The appeasement of the northeastern hill tribes was also necessary for the British policy in Assam, under which the Brahmaputra trade route might come under the commercial control of the East India Company. The early Assam kings regularly traded minerals and such forest products as aloe, sandalwood oil, and paper with the neighboring tribal states and with Tibet, Burma, and China. Elephants were captured in the forests of the Duars Plain for use in the royal armies and were sold to neighboring kingdoms in exchange for salt and silver. These prosperous enterprises attracted the attentions of the East India Company. As early as 1787, the British sent a superintendent into the region to oversee their interests there and, they hoped, to capture for the British coffers some of the gold, precious stones, sandalwood, perfumes, textiles, elephants, and other trade items that came down the valley from the north.

In 1817, the Burmese advanced into the valley of the Brahmaputra River. It was a short-lived invasion. In 1825 the Burmese were expelled by the British after a yearlong war. Thereafter, the British began to administer Assam in earnest, partly to collect revenue to recoup the losses they incurred during the Burmese War, estimated at over £5 million sterling, but also to show a profit in the Calcutta ledger books.

The earnings from trade was an important revenue source in the northeastern range but not the only one. The British also imported opium into Assam from their production areas located elsewhere in the colonies. British policy encouraged the use of opium among the northeastern Himalayan hill people. Sir Andrew Mills, in his official report of 1854, wrote, "Opium they should have, but to get it they should be made to work for it."[9] The promise of addiction thus was the reward for toiling for the Queen of England.

In the central Himalayan region, the old caravan routes moved salt from Tibet to India and surplus products from Nepal to both the neighboring countries. Much of the Nepal-Tibet caravan exchange was small-scale border trade in subsistence goods. Considerable amounts of commercial trade occurred, however, between Nepal and India. In particular, agricultural products from Nepal's tarai, especially rice, were shipped almost exclusively to northern India, especially to the city of Patna, where they were sold for badly needed cash.

The tarai forests also supplied timber, medicinal plants, and wildlife such as hawks and elephants. Villages provided finished textiles and wood carvings. The hill region of Nepal grew cotton and cardamom. All those items ended up in the trade business. Furthermore, copper was mined in Nepal in the high mountains and sold to markets in northern India. By 1806, most of the copper used in the valley of the upper Ganges River came from the mines in central and western Nepal.[10] It was used for coinage and for making vessels. The Nepal trade was directed by rulers based in Kathmandu who worked through powerful rural agents. It moved through the market towns in the countryside, called *mandis*, which were, in turn, controlled by the government-appointed tax collectors. In the eastern range, the trade was diverted through regional centers called *golas.* In either case, Nepal's system of customs and taxation organized the Himalayan trade so that goods flowed out of the highlands and onto the plains, rather than between places located in the mountains. That geographical pattern of trade still predominates in Nepal.

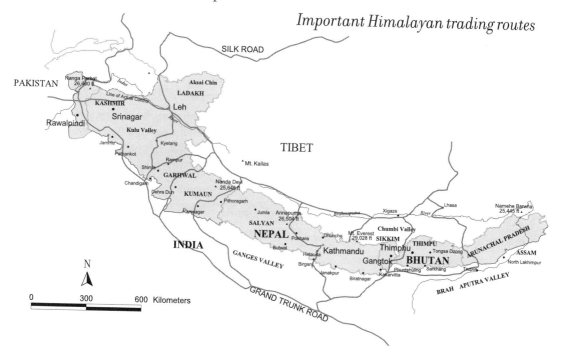

Important Himalayan trading routes

A caravan of porters travels the route between Nepal and Sikkim. Photo by P. P. Karan

The British worked through the Indian traders and Nepalese brokers to gain favorable trade concessions for the East India Company. They had no formal access to Nepal, but they hoped to gain a monopoly on the central trade routes by placing intermediaries close to the action.[11] The British commercial agent Brian Hodgeson, for example, spent most of his life in Kathmandu, monitoring the Himalayan trade and determining how it might be diverted for the benefit of the colony. He was concerned foremost with how the trade routes might carry British-made products to Tibet and China. He foresaw English broadcloth in Tibet, linen in Nepal, and velvet in China. He reckoned that there would be a strong demand in the Himalaya for British crockery and glassware, for mirrors and chandeliers, and for fine chintzes.

Hodgeson was obviously wrong in much of this, but he nonetheless accurately predicted that British commercial interests in the mountains could be fortified by favorable commercial treaties with Nepalese rulers. These would be forthcoming with the rise of Rana rule in Nepal in the mid-nineteenth century. The Rana aristocracy conveniently saw that its own political future could be best insured by alliances with the British. Under the appraisals of men like Hodgeson, British colonialists understood the possibilities of lucrative trade with Tibet, which paid gold bullion for commodities such as rice, cotton, and tobacco. The Himalayan trade routes could also be the conduit for tea and wool from China and Tibet, which were in great demand in both India and Europe. The Ranas, in their turn, saw that their own prospects could be improved by the patronage of the British colonists. Hodgeson and the other British commercial agents were patriots. They did not stop with their ambitious plans for hegemony over the Himalayan trade. They wanted to take over the mountains. These colonial agents proposed that it would be a good thing for the British to settle in the Himalaya, to establish residencies there, and from those headquarters to measure and extract the abundant natural wealth of the mountains. Their motive was commercial profit, and they were the provocateurs by which the wealth of the central range would be diverted to the imperial system of commerce that extended worldwide.[12]

In the northwestern region, meanwhile, the British were greatly motivated by the well-established trade that already moved along the old Silk Route between Asia and the Mediterranean. The ancient caravans began in Turkestan and led all the way to Rome, crossing en route the vast empires and frontiers of Afghanistan and Persia. The Silk Route carried a brisk trade of silk and lacquerware from China and textiles, rugs, spices, and jewels from central Asia and India. Yak caravans transported silk, jade, and salt across the Aksai Chin Desert north of Ladakh. Slaves, wool, and horses were traded between Kashmir and Kashgar. Tea made its way from Canton via India and the western Himalayan passes to Tashkent.

Considering all this potential wealth, the British sought to control some of it by acquiring large sections of the northwest mountains and by restricting trans-Himalayan trade to those routes that passed through British territory. They were not alone in such designs. The Russians and Chinese had similar interests in the land north of the Great Himalaya. The Russians built roads and railway lines south across the Kirghiz steppes to Tashkent and to the Afghan border in order to gain access to commercial regions. The Han Chinese overlords living in Xinjiang tightened their grips on eastern Tibet, Kashgar, and the remote Aksai Chin Desert in order to bolster their authority in the western ranges.

Against such maneuvers, the British sought ways to control the mountain passes for the dual purpose of gaining access to the central Asian trade and fortifying their Indian border against the possible aggressions of either the Russians or the Chinese.[13] The strategic conflicts over trade, access, and border security that began in the region during the colonial period continue to trouble the relations today between Pakistan, India, and China. Numerous wars between the neighboring Himalayan countries have since been fought over the placement of the boundaries, and the border region remains a politically sensitive area for all three countries.

COLONIAL EYE ON HIMALAYAN RESOURCES

The commercial interests of the British did not stop with a command over the Himalayan trade that passed through their domain. It extended also to the wealth that might be gained from directly exploiting the natural resources of the mountains or from collecting revenue on the agricultural production. In regard to natural resources, the British focused mainly on timber products, which eventually became their primary commercial concern in the western Himalayan forests. In the eastern range, meanwhile, the attention of the colonialists was on developing the tea gardens of Darjeeling, Sikkim, and Assam.

When the British gained control over the Punjab hills in 1849, they redrew the administrative map of the mountains. In some cases, the local rajas were provided nominal authority in running the affairs of their petty kingdoms, but elsewhere the British assumed complete and direct supervision. The colonial interests were essentially twofold: to expand market agriculture and to exploit the mountain forests. Despite some limited attempts to promote the planting of temperate fruits and to increase grain production for export, capitalist farming never really took hold in the western Himalaya during the British reign. The forests of Kashmir, Garhwal, and Kumaun, on the other hand, became a major source of timber products for the colonial marketplace and a primary source of income for British entrepreneurs.

The environmental historian Richard Tucker has shown how the mountain forests were extensively cut as early as the 1770s in order to supply growing colonial demand for fuel wood, railroad materials, and construction timber. The ubiquitous lowland sal (*Shorea robustus*) forests, found all across the Himalaya, provided great quantities of versatile hardwood; the mid-elevation chir pine (*Pinus roxburghii*) forests were tapped for resin; and the highland conifers, including *deodar* cedar (*Cedrus deodara*), supplied lumber for the regional timber markets.

Estimates of timber extractions across the region during the colonial period vary widely, but the records of the British administration provide the basis for some reasonable statistics. In the Dehra Dun district in the Garhwal Himalaya, the average annual export of timber to the plains in the early 1900s is estimated to have been over 6,000 cubic meters, while forest cutting for fuel wood and charcoal was about 27,000 cubic meters. In nearby Kumaun province, meanwhile, the timber production amounted to over 300,000 cubic meters, and fuel-wood production was as high as 700,000 cubic meters.[14] In order to maintain such large forest outputs and to insure the steady supply of timber for public works projects, the British introduced western silviculture into the region. Their goal was to maximize specialized tree production for export, but this was at the expense of the diverse forest needs of the local inhabitants.

The cumulative effect of British forest policies was to create a colonial monopoly over much of the western Himalayan woodlands. The Indian Forest Act of 1865 established the responsibilities of the Forest Department and gave it jurisdiction over forest management. A key factor in the colonial appropriation of mountain timber in the early decades of the twentieth century was the establishment of new forest classifications. The New Reserves, as they were called, provided the Forest Department with exclusive revenue rights in the designated forest areas. Petty indigenous traders were effectively excluded from the new timber markets, and villagers were not allowed to cut trees, graze animals, cultivate, or collect minor forest products in the New Reserves. The prohibitions curtailed the options of villagers in how they could use the forests, and suppressed the local management of environmental resources.

Those actions led to numerous forest-related disputes and protests.[15] Opposition to the New Reserves took the form of litigation, demonstrations, and arson. In 1921, the Forest Department reported over 830 square kilometers of Reserve Forests in Kumaun and Garh-

Tea plantations in Darjeeling, India. Tea was introduced to this area in the late 1800s by the British colonial planters. It remains an important source of revenue for the Darjeeling region. Photo by D. Zurick

wal destroyed by fire. Peasant uprisings against the impositions of colonial forest policies led to a series of special Forest Commissions and to the eventual cancellation of the New Reserves in 1922, which henceforth were administered by the Revenue Department simply as State Forests.

Ecological and social damages, though, were already pronounced by the time of the policy changes. In the places where overharvesting and arson had occurred, extensive mountain landscapes were left barren of trees. British silviculture practices provided in some cases adequate tree plantings, but they did not produce the entire range of forest habitats that rendered the subsistence products needed by the local people. The grazing restrictions imposed on the semi-nomadic Gaddi herders who inhabit the western mountains by the forest policies led to serious degradations in the pasture lands because movement of livestock was limited to smaller, more heavily regulated territories. Finally, indigenous forest management systems and the customary land rights of the native people were viewed as antiquarian, not practical under the new commercial demands placed on the forests by the colonial rule, and therefore were abandoned. The results of such events were not only a deterioration in the quality of the forests but also a diminished capacity of local people to manage them.

In the eastern Himalaya, efforts to exploit the abundant forest wealth were curtailed by the fact that most of the forests were inaccessible for commercial purposes. Rugged topography and lack of roads combined to curtail the harvests of exportable timber. There were no sizable state-held forest reserves in the Assam Mountains until the 1920s, and even then the forests were not well surveyed for commercial logging purposes. Consequently, commercial timber production in the northeastern Himalaya remained low until the middle of the twentieth century.

Expansion of shifting cultivation in the region, however, posed a considerable threat to the forested areas. A stream of immigrant farmers from the Brahmaputra valley entered the uplands in the late 1800s and early 1900s, which led to additional pressures on the forest lands. In World War II, when commercial timber production increased in Assam as a result of military demands for timber and the placement of new roads in the region, the added burden of commercial forest cutting finally advanced into the remote mountains.

Throughout much of the colonial period, the British focused their attentions in the eastern sectors on the cultivation of tea plantations. The proximity of Darjeeling to Calcutta and the ports of the Bay of Bengal facilitated the shipment of tea to Europe, making it a lucrative proposition for traders. When it became legal for Europeans to own land in India after 1833, planters bought up huge tracts of forested hill land in Darjeeling and in western Assam. In 1854, the government began providing direct incentives to the settlers to clear the land by awarding prime acreage to Europeans willing to plant tea gardens. By 1871, the tea planters owned 700,000 acres in the mountains, distributed among 300 separate tea estates. Plantation workers for the Darjeeling estates were recruited from immigrant farmers streaming into the region from Nepal. Farther to the east, in Assam, Bihari people from the plains were employed on the new tea plantations. The workers often toiled under the appalling conditions of indentured labor.[16]

The establishment of tea estates in the eastern Himalaya expedited the colonial penetration into the mountains. As the tea gardens expanded, forest areas diminished. Land for tea bushes and wood for packing crates put heavy demands on the accessible forests. The new plantations encroached on village common lands, and usurped the land rights of indigenous residents. To secure the tea-growing areas against the insurgencies of the mountain tribes, the colonial military was heavily garrisoned in the mountains. The economic value of the region rose with such developments in the estimation of the British, and consequently so too did their concerns about the interlinked issues of border security and territorial control.

The tea plantations in the eastern Himalaya were a premier example of the political and ecological imperialism practiced by the colonists. When the British finally relinquished their colonial rule in southern Asia in 1947, foreign designs on the wealth and territory of the mountains did not end. The struggle for hegemony there continued under the post-Independence efforts of India and Pakistan. In many ways it even accelerated as the independent nations sought to define the territory newly under their control and to reap immediate benefits from it. As in the past, the native people of the Himalaya continued to bear the cost of the new postcolonial efforts to exploit the wealth of the mountains.

In the regions of the Himalaya that were not under the direct control of the British, such as Nepal and Bhutan, the ruling autocracies ran their independent kingdoms much as feudal estates. The Rana oligarchy in Nepal, for example, administered the rural areas through exploitative arrangements based on nepotism and patronage. High-caste Hindus gradually gained almost total dominance in the mountain economies, which then became oriented toward the southern markets. In eastern Nepal, many people escaped the difficult conditions by fleeing across the border to Sikkim and Bhutan, creating in those places a large Nepalese population. In Bhutan, the nobility, vested in the Wangchuk dynasty since 1907, maintained exclusive access to the country's resource wealth and political power. Much of the surplus production from agriculture went to maintaining large monasteries and to creating a sense of civic order. Whereas Bhutan's rulers sought to insulate the country's populace from foreign influences, they nonetheless secured their own positions by establishing personal relations with foreign powers. After India achieved its independence and the Chinese occupied Tibet, Bhutan forged a close alliance with India to safeguard its borders against Chinese aggression, thereby opening itself to more lowland control.

A view of mountain territory in the Great Himalaya zone of Nepal. Photo by P. P. Karan

A Divided Geography

THE MODERN BORDERS OF the Himalayan countries derive in part from the earlier colonial divisions that were made for purposes of political control, trade, and access to resources. They also reflect the more recent political ambitions of the independent Himalayan nations. The partition of southern Asia in 1947 split the line of control in the western Himalaya between Pakistan and India. Since then, the two countries have skirmished frequently over disputed territories in Kashmir and Ladakh. Pakistan currently controls about one-third of Kashmir, known as Azad ("free") Kashmir, and India controls the remainder; their respective territories are separated by the 480-mile Line of Actual Control.

The border wars between China and India have centered on the Aksai Chin Desert, north of the Indus River, and culminated in the Sino-Indian War of 1962. That war was resolved when a fixed boundary line was established north of Ladakh. Nepal has served historically as a buffer between India and China in the central Himalaya, so the border there has remained basically the same since the Gorkha period. In the eastern sector, the two superpowers clashed over the placement of the McMahon Line, which failed to adequately demarcate the international boundary between Arunachal Pradesh and the Kham region of Chinese-administered Tibet. The policy of both India and China in these areas has been to increase military access to the border region by building roads into the mountains and by placing troops in large numbers along the frontier posts. The military presence has contributed to a high level of tension in many of the frontier regions.

The proximity of Nepal, Sikkim, and Bhutan to the territorial disputes between India and China has largely determined their respective roles in the strategic conduct of the Himalayan borderlands. Sikkim became a protectorate of India under the terms of a treaty negotiated between the two countries in 1950. It was formally annexed by India in 1975, in part to resolve internal political and economic problems but also to bolster India's line of defense along the strategic Chumbi Valley in Tibet. Sikkim remains a sensitive area for both India and China, and China continues to refute its annexation by India.

The territorial integrity of Bhutan against possible Chinese aggression became a paramount security issue for both India and Bhutan in the 1950s, when a series of Chinese maps showed significant Bhutanese territory marked under Chinese control. Despite assurances of the Chinese authorities that the maps were simply cartographic errors, their continued publication caused enough worry in Bhutan for it to draw closer to its southern neighbor and to seek support from India in the event of frontier aggressions by the Chinese. In effect, this meant an important change in Bhutan's policy, which formerly had relied upon seclusion for its defense. Bhutan closed its border with Tibet and accepted India's offer of trade and transit relations.

Wedged in between China and India, Nepal has based its foreign policy on its pivotal location. Since the mid-1950s, when both India and China began providing foreign assistance to Nepal, the kingdom has tried to capitalize on the risky strategy of working the territorial interests of the two superpowers to its own benefit. This has largely

A newly opened military road provides access to remote areas of the western Himalaya. Such roads cross the famous Inner Line, which demarcates the frontier zone of India, and are maintained by the Border Roads Organization. Photo by D. Zurick

paid off in terms of infrastructure development, since both China and India have invested heavily in road building in Nepal. The routings have contributed to Nepal's modest achievements in economic development. But Nepal's strategy is flawed, because it has made the country's domestic affairs overly dependent upon the larger security concerns of its two neighbors. This is exacerbated by the fact that Nepal is a landlocked country and must conduct its foreign trade primarily through India.

To a large extent, India's and China's claims are authorized by the roads that have been built in the mountains since the 1950s, most of them for the purpose of military access. The 1962 border war between India and China, for example, compelled India to build over 11,000 kilometers of roads in the western Himalaya alone. Those routes are now open to the public on a restricted basis, through the issuance of Inner Line Permits, and connect once remote places in Zanzkar, Spiti, and Ladakh with the rest of the country. Passage on the military roads is now possible for both Indians and foreigners, although the travel is closely monitored by the India Border Roads Organization. The recent lifting of restrictions on travel along the Indian frontier roads is meant to promote greater integration of border communities with the national society and economy. Pakistan made similar efforts at opening its northern frontier agencies when, in 1973, it jointly constructed with China the Karakoram Highway along the Indus River. Islamabad and Beijing were thus connected by road, and the distant valleys of Hunza, Kohestan, and Balistan came under the sphere of direct Pakistani rule. The Karakoram Highway was completed in the mid-1980s. It has since had enormous impacts on the society and environment of the Himalayan valleys located in the shadow of Nanga Parbat.[1]

The development of the Nepalese infrastructure clearly reveals the security interests of its neighbors. The Indian army built the 192-kilometer Tribhuvan Rajpath in 1956, connecting Kathmandu south to the Indian border. The 114-kilometer Arniko Rajmarga, constructed by China, was completed in 1967 to connect Kathmandu north to the Chinese border near the town of Kodari. The 200-kilometer Prithvi Rajmarga was also built by the Chinese, to connect Kathmandu with the Pokhara Valley. The 1,050-kilometer Mahendra Rajmarga, also called the King's Highway, which connects the eastern and western sectors of the tarai, was built by the combined efforts of Britain, the Soviet Union, and India. More recent efforts at road building in Nepal, financed by an alliance of international donors, focus on feeder roads that link rural mountain areas with the tarai highways. These latter efforts reflect Nepal's interests in enhancing the development potential of its interior regions, rather than the strategic goals of India or China.

The modern territorial interests and infrastructural developments of the Himalayan countries reflect their desire to develop the resource potential of the mountains for the purpose of national economic growth. In the border areas where territory is disputed or where central authority over the landscape is otherwise not secured, national development is hampered. Hence, the regional interests of the Himalayan countries include extending political and administrative control to these remote places. Throughout the mountains, the advancement of national development is enhanced by the increased accessibility of the mountain regions, which also more tightly integrates rural populations with the larger concerns of the state. As so often happens, the strategy fails where it does not include the means of promoting local economic advancement; in such cases, the new roads simply provide alternative routes for people and resources to leave the mountains.[2] Nonetheless, the conventional view of policy makers is that the natural resources of the mountains can be exploited best for purposes of economic growth when roads and markets are built into the highlands.

The consolidation of mountain areas under the political authorities of the central Himalayan countries has encouraged new forms of spatial development in the mountains. In Nepal and Bhutan, for example, a series of five-year plans initiated since the early 1960s have organized the development of rural areas according to broad spatial frameworks. In Nepal during the 1970s and 1980s, those schemes were known as the Integrated Rural Development Projects and were the primary focus of national development. The large area, multisectoral programs were conceived at the national planning level and implemented among numerous gateway towns in order to facilitate economic change. The basic goal was to organize mountain land and resources for national development.

But the fact that the design of such projects included little local input diminished their overall capacity for equitable development in rural places. Instead, they often simply usurped local systems of environmental management, allowing local elites to enhance their economic positions in the villages. Most important, by incorporating the villages into the national economy, the development programs have led to a breakdown of cultural autonomy and to the loss of environmental resiliency among many communities. Those changes, in turn, have led to new problems for resource management and equitable social change.

The primary focus of Bhutan's initial five-year plans was to enhance the capacity of the mountain areas to produce export resources in order to achieve national economic self-sufficiency. It was determined that this could be managed best by developing industrial export programs that targeted the nation's forests and minerals. Most of those resources are located in places that are not easily accessible, thus making their economic exploitation difficult. Some coal, limestone, marble, and slate deposits have been mined for domestic industrial use or exported abroad in order to generate foreign revenue. Commercial timber extraction in Bhutan increased dramatically during the 1980s, though, from 77,000 cubic meters in 1981 to 235,000 cubic meters in 1986. During that same period, Bhutan's forestry exports to India increased in value by threefold.[3]

The current national plans of Bhutan strike a somewhat different note. The new programs also seek to mobilize Bhutan's resource potential, but they apply more caution to the expansion of the public and private sectors. The revised goal of Bhutan's plans is to preserve natural habitats and indigenous cultures while still achieving measurable economic growth. Bhutan has made it clear in the most recent development reports that the country's economic future must be compatible with maintaining its cultural traditions and natural environments. To the extent that Bhutan's policies protect the territorial rights of its indigenous residents, they constitute dramatic forward-looking approaches to the region's environmental and social dilemmas. In many ways, Bhutan has become a model for sustainable development throughout the range.

Territorial development in the western Himalaya, meanwhile, has followed a path determined largely by the economic interests of lowland India. Those interests have focused mainly on the combined goals of increasing industrial output in the mountains, extracting its forest and water wealth, expanding the base of commercial farming, and most recently, promoting tourism in the mountains. In many, if not most cases, the beneficiaries of these territorial policies are the industrial societies of the distant plains, the local political elites, or the commercial entrepreneurs who serve as vital links between the highland resource systems and the lowland markets. Only recently have the development agendas in India come to incorporate a greater sensitivity to the fragile conditions of the mountain environment and a nominal commitment to meeting the basic needs of mountain residents. The implementation of sustainable development programs, however, falls far short of their design, and have so far failed to achieve much of an equitable balance between the subsistence resource needs of mountain people and the commercial prospects for national growth.

Timber-production programs initiated by the British have been carried forward by independent India where state control over forest lands allows politicians and businessmen to exploit the public forests for both public and private gain. In Garhwal, the timber merchants concentrate their efforts around Uttarkashi in the district of Tehri; and in the Doon Valley, near the town of Dehra Dun. The commercial bias of forest management in the mountains around the Doon Valley is particularly outstanding and, in combination with widespread limestone quarrying, has rendered a seriously degraded natural environment in the valley and among the surrounding hills.[4] Houseboat tourism in Kashmir, meanwhile, has contributed to high pollution levels in the Dal Lake. Elsewhere in the western region, water development schemes for hydropower are widespread and contribute additional territorial pressures as the dams, reservoirs, and related infrastructures encroach on the native forests, grazing lands, and subsistence agricultural areas.

All across the Himalaya, national efforts to develop mountain resources for economic growth ignore the long-standing claims of native residents and frequently conflict with local territorial needs. When the rural communities become alienated from their lands, a common result is the overall impoverishment of villagers and their migration to other areas, including the rapidly growing hill towns. In some instances, local communities benefit from strong leadership and have either resisted the demands of the lowlanders or forced the adoption of more sustainable kinds of economic development. We find examples of environmental resistance among communities across the Himalaya over such things as the placement of large dams or the large-scale commercial extraction of resources. Where such encroachments are extreme, and community politics are strong, conditions exist for prolonged resistance struggles.

Environmental Activism in the Mountains

The fixing of national boundaries and the development of territory-based economic programs presume a political order in the Himalaya that may not really exist. What we often find in the mountains is a polyglot of ethnic and territorial affiliations. Where the environmental security of local communities is threatened by the territorial claims of national powers, disputes about resources and geography regularly ensue. Examples of communities asserting their authority over environmental matters are numerous and include the separatist movements in Uttarkhand, Gorkhaland (Darjeeling area), and Bodoland; the opposition to hydroelectric dams in Tehri Garhwal, Nepal, and Sikkim; and a number of comprehensive environmental movements such as the well-known Chipko Protest.

A history of isolation in the eastern Himalaya has produced there a diverse assembly of tribes and communities that have little allegiance to the Indian government. The eastern tribes are marginalized and located at a far distance from the political center in Delhi. They agitate frequently for cultural identity. The interests of India in Arunachal Pradesh are primarily related to border security issues, to the fear of a loss of national territory to the insurgents, as well as to a desire to exploit the abundant timber, water, and mineral wealth that is believed to be located there. Recent migrations of people from the plains into the eastern mountains compound the problems of land and resource conflicts in the highland communities. As the pressures on their local environment have increased, eastern Himalayan communities have developed even deeper resentments toward the central government, which they increasingly see as a matter of new colonial behavior.

In the western Himalaya, a similar situation unfolds. The well-known separatist agitation in Kashmir is linked to religious affiliation and to geographical problems between Pakistan and India, but it also is a matter of political representation, domestic territorial claims, and local sovereignty. The mountain districts of the state of Uttar Pradesh, located east of Kashmir, compose the 45,000-square-kilometer territory called Uttarkhand. The uneven politics of the Uttarkhand region reflect its minority status in a state that is dominated by a plains-based population. They also reflect the fact that the natural resources of Uttarkhand have been exploited historically for the benefit of the lowlanders. The struggle of people in Uttarkhand for self-determination and the desire of people in the plains for the resource wealth of the mountains have established the conditions for long-term regional conflict. The result is an active, intermittently violent struggle for sovereignty and separatism in Uttarkhand.

Two important rallying points for the territorial struggles in the Garhwal region are community opposition to the construction of the Tehri Dam and the forest-resource–based Chipko Protest. The Tehri Dam was formally approved by the Uttar Pradesh government in 1976. It was to be situated on the Bhagirathi River above the village of Tehri. The dam, itself, would be 260 meters high, and its reservoir would extend forty-four kilometers up the Bhagirathi valley. Numerous villages and extensive areas of farmland would be submerged. In 1978, formal opposition against the dam was organized by activists living and working in the Tehri community.

The long drawn-out battle in Tehri centers around four key issues: the seismic hazards of the region; the eventual submersion of numerous villages, agricultural land, and pilgrimage routes; the forced relocation of over 70,000 people; and the lack of any significant local benefits from the project. Those opposing the dam see it as yet another example of the Himalayan communities bearing the brunt of social and environmental displacements so that the plains-based societies may benefit. A series of protests in the late 1980s and early 1990s drew international attention to the Tehri project and halted for a time construction work at the dam site. The future of the Tehri Dam remains uncertain, although the construction now continues, albeit amid frequent disruptions.

The concerns of the Tehri opponents over the loss of territorial authority are echoed elsewhere in the mountains of Garhwal and in neighboring Kumaun, where the Chipko activists have united many communities against timber merchants and political bosses who seek to exploit the public forests for private economic gain. The history of forest protests in Uttarkhand began early on, when the national state claimed jurisdiction over forests that were traditionally part of the village common property. The Chipko Protest is the latest in a series of village protests over government encroachment on community land rights.[5] Throughout the 1970s and 1980s, several large demonstrations, led mainly by village women, took place against the activities of timber merchants in Uttarkashi, Chamoli, Nainital, and other nearby hill towns. The goal of the protests was to publicize the fact that timber and charcoal contractors were destroying the forests and thereby causing the problems of mountain deforestation for which the hill people were conveniently blamed.

Chipko literally means "to hug," and it refers to the highly publicized actions of village women who, with locked arms, surrounded individual trees in an effort to block the work of timber cutters. It has since become equated with a more comprehensive nonviolent movement to protect the mountain environment from the devastations of commercial resource merchants. What began as a village effort became a national movement that subsequently gained international attention. Factions eventually formed within the group as it became more highly politicized, and the strength of the movement has diminished in recent years. Nonetheless, the ideals of the Chipko Protest still inform environmental movements in various parts of the Himalaya.[6]

The resistance struggles tend to be grassroots responses to development projects that threaten the environmental security of the local villages. They position local communities squarely in the middle of resource rights and responsibilities. They also contribute to a more comprehensive understanding of the environmental dilemmas that face the region, for they show a history of environmental change that is tied to political, as well as economic, control over resources. In its modern setting, the Himalayan region is characterized as a place of great ecological instability. There are many ideas about the cause of environmental decline. Some observers attribute it primarily to the demands of the growing mountain population for food, fodder, and fuelwood. Others link it to the historical inequities of expanding commercial economies and political control. A third group suggests that the environmental problems in the Himalaya are mainly of natural derivation, owing to the dynamic conditions of the mountain environment. In our estimation, it is a combination of all three factors.

The interpretations of land degradation also argue that the high levels of resource degradation in the mountains have a direct and adverse impact on the quality of life all across the plains of Pakistan, India, and Bangladesh. They point to a vast region that is seriously out of balance, where the stripping of the mountain forests may have led to greater floods in the Ganges Plain. Others refute that idea and discount the role of the highlands in producing the degraded conditions of the lowland plains. Finally, a number of observers simply acquiesce to the uncertainty that surrounds the problem of environmental change in the Himalaya and forge ahead with new proposals for development. In all cases, the sustained quality of life in the mountains is tenuous. The risk of living in the Himalaya increases with the passage of time.

PART THREE

a question of balance

Sediment loads in the Sun Kosi River near Bahrabise, Nepal, show the effects of soil erosion from both human activity and natural geological processes.
Photo by P. P. Karan

A Chain of Explanation

8

each year on the evening of the full moon, in the springtime month of Baisekh, a small group of elderly village priests climbs through the oak forests above Phalabang in western Nepal. Their steep walk takes them to the summit of Mount Tharkot, where a crumbling wall of stone encloses a grassy platform. In the center of the enclosure the priests place a copper vessel of water carried up from the spring at Chaite Dada. Along the leeward side of the wall, protected from the wind, they arrange pine boughs and sleeping mats. The priests will spend the next two days at the spot in purification rituals, and they will pray for rain.

Looking east and west, the priests can see the undulating and broken crest of the Mahabharat range, the southern ramparts of the Nepalese Himalaya, which stretches to the horizon in both directions. Fifty kilometers to the south is the Ganges plain of India, lost from view in a haze of dust borne by the pre-monsoon winds. In a northward sweep, the middle mountains extend some thirty-five kilometers toward Tibet. They appear as a succession of ever-higher ridges that dissolve into the snowy summits of the Great Himalaya.

From atop the windy, sun-drenched peak the priests view their close surroundings. They see the tidy clusters of farmhouses that dot the ridges and lower valleys. The sloping corn fields and small kitchen gardens interspersed between the homes produce food for village families. The priests see the rice terraces that cascade far below, among the bends of the Luwam and Sarda rivers. Their wet surfaces sparkle in the low sun. They watch the village livestock graze in the upper pastures near the rhododendron forest and listen to the small sound of cowbells as the animals move lazily among the trees. It is an intimate and familiar scene. The priests' bond with the mountains is strong, for they are farmers as well as religious men and have carved their lives and homes from the fragile mountainsides.[1]

From afar, the life and land of Phalabang appear idyllic, even enchanting. They express a certain easy gracefulness. By their work and close attentions to the land, the Phalabang villagers have rendered a landscape that has both an aesthetic appeal and a productivity that is unique on earth, one that is highly expressive of a sophisticated knowledge about nature and ecological principles. But their world is a dynamic one, where the loss of forests, soil erosion, and river siltation are the visible signs of environmental damage. According to many commentators, such problems threaten natural systems throughout the Himalaya, and the human societies that are dependent upon them suffer as a result. The Phalabang villagers, of course, know a great deal about such things; they have lived with them for a long time.

In the past few decades, scientists and policy makers in the Himalaya have come to better understand the nature of the difficult conditions presented by the natural systems for human life in the mountains. Some of the problems in the region, such as unequal ownership of land or limited access to resources, are linked to the region's imperial past and to the historical policies that extracted the natural wealth of the mountains for the benefit of outside rulers. Other problems, however, reflect more current conditions of society and economy in the mountains. These problems derive not only from the basic needs of the mountain villages but also from the spiraling demands of national development and of a global economy that has penetrated even the most sequestered mountain places. Under the circumstances of local life and the wider imperatives of modernity, there is no simple answer for the troubled conditions that now beset many places in the mountains.

One of the main elements of Himalayan land degradation is the linkage between the demands of growing numbers of people and the fragility of the mountain resource base. The population is increasing at an annual rate that exceeds 2 percent, resulting in a doubling time of less than thirty-five years. Even the most remote alpine regions now support numerous farming villages and migratory herders. In some cases, such as that of the populous middle-mountains zone, more than 1,000 persons may occupy every square kilometer of arable land. Those places are among the most densely settled spots on earth. In such localities, the pressure of population on the land is a major element in the mix of factors that lead to environmental decline. It combines with other factors, such as commercial resource development, to surpass at times the natural carrying capacities of the land.

Under these conditions of dense human settlement, the villagers may spend much of a day walking to the forests to gather a supply of fuel wood. Women may have to wait for several hours each morning at the village spring, waiting for the slow seep of water to fill their vessels. In most Himalayan localities, all the available farmland has been taken. On the steepest slopes, the land is carved into stepwise terraces for level surfaces on which to grow crops. The distribution of precious water through the handhewn canals and gates of the traditional irrigation systems also depends upon the careful maintenance of the terraces. When it rains hard and some of the fertile topsoil washes down the sloping fields, the farmers carefully gather what dirt can be saved and carry it back to the upper parts of their fields. The continuous rearrangement of soil is a tedious and unrelenting business for them.

The farmers in Phalabang, and in countless other villages like it, inhabit a world in which every act is measured against the needs of the household and the capacity of the natural environment. Their survival requires paying close attention to both, and especially to their changing circumstances. Living so close to the land, abiding in it, is the foundation of rural society and the crux of the ecological and human dilemmas that confront life in the Himalaya Mountains. It is a delicate balance.

The villagers have inherited a landscape that is the product of a long history of human settlement and use. The medieval period of exploitive princes and kings, the colonial era of imperial demands on mountain resources, and the modern decades of national development have taken a heavy toll on the Himalayan habitats. The natural resources in the mountains are now reckoned by price. Profit rather than subsistence has come to largely determine the nature of rights to the land. The entrepreneurship that is meant to bolster national and personal wealth also portends great demands on the natural capacity of the mountain ecosystems.

The quiet of some villages is disrupted now by dynamite blasts of the road crews, by the whine of the timber sawmills, by the low hum of the new water-powered generators. Horizons may be punctured by satellite dishes or power cables. Hotels, administrative offices, and commercial establishments straddle many of the main travel routes in the mountains, displacing the old resting places shaded by sacred ficus trees. Such features are commonplace nowadays in the Himalaya, whereas only a few decades ago they were still novel additions to the landscape. These innovations do not obliterate the villages; they do, however, intrude upon them and juxtapose rural life against the wider demands of the industrial world. They add new, ever more complicated layers to the history of social pressures on the Himalayan environment.

As is true elsewhere in the Himalaya, the Phalabang villagers still rely almost exclusively on firewood for cooking and for heating their homes in the winter. Every household burns more than an armload of wood a day—several hundred kilograms each year. The measurement of fuel-wood consumption is a common component in many environmental analyses, but it is misleading to reduce such usage to the simple equivalent of energy equations. The central feature of all Himalayan homes is the hearth. It is usually a molded clay oven or an open pit fire. There, the household members gather, to huddle for warmth in the cold, to prepare their daily meals, and to share in the simple celebrations of family life. The windowless dwellings of Phalabang are brightened by the glow of the fires. The hearth makes the home.

But the supply of fuel wood in Phalabang is more difficult to come by since the forests have shrunk, and the security of the home may therefore be diminished. The environmental reports of the 1970s and early 1980s, written by such agencies as the World Bank, predicted that the ancient forests of Nepal would be gone in a matter of only fifty years because of fuel-wood cutting, and as a result the mountains would turn into a giant, muddy desert. Fortunately, that will probably not come to pass. The actual conditions of the Himalayan landscapes are not as bad as those painted in the earlier predictions. Nonetheless, within the lifetimes of many villagers, the canopy forests in Phalabang have dwindled by a third; places that appeared as a healthy forest in aerial photographs taken in 1953 show now as disturbed areas of scattered trees, shrubs, or grassland. The time needed to gather fuel wood has increased as the trees recede farther from the settlements. The expansion of hillside gullies has created a sort of badlands topography that cuts into some of the best farmland, reducing the productivity of those areas. The villagers in Phalabang complain about the lack of fodder for livestock and the diminished flow of water in the village springs. They link those problems to their perception that the land is, in their manner of speaking, "drying up."

In their attempts to reverse such trends, the Phalabang residents established in the mid-1980s a tree nursery in the village, marked several areas of public lands for tree planting, and restricted grass cutting and livestock grazing in the most degraded common lands. Their intention was to reclaim the old village forests and to make eroded agricultural lands productive once again. It was time, they believed, to heal the wounds in the land. Their efforts embraced solid conservation practices and extended the villagers' environmental traditions into the modern age, thus benefiting from both the wisdom of the past and the new scientific innovations.

Traditional farming in the Himalaya, upon which most villagers in Phalabang depend, requires the careful cycling of nutrients from forest to farm field. Most days in Phalabang the household livestock are taken to graze in the pasture; in the evening the animals are bedded down in stalls and their dung is collected for composting. The large numbers of livestock kept in the village provide the valuable manure that is applied to the fields in order to replenish the nutrients in the exhausted soil. The animal dung is mixed with dry plant cuttings and piled into small mounds on the fields. When rain appears imminent, the villagers set fire to the stacks of damp compost. The smouldering piles are then spread onto the land and left for the night. Early the following morning, the farmers plough the fields with strong teams of bullock. As the compost is turned into the ground, the circle is closed. The forests and farms of Phalabang are joined under the ploughshares and the falling rain. When the priests of Phalabang view the land from their vantage point atop Mount Tharkot, they see in it all the possibilities of village life. They see in the landscape below them the work of humankind and the gifts of nature. They see the neat fields of maize, smell the smoke drifting upward through thatched roofs, hear the crow of roosters and the squeals of laughing children. They see where the forests have been cleared and where landslides now erode the land. They see the new plantations of young trees. Across the river, they watch where the new road winds up the Sarda valley toward the market town of Salyan. It is a mixed scene, filled with contradictions, with worry and hope, but the village remains the center of rural life, sustained as always by the productivity of the mountain landscape and the work of the farmers.

One finds in such places an extraordinary balance between the providence of the land and the needs of the family, the claims of the nurturing hearth and the wider engagements of the outside world. Those relations, so common in Himalayan villages, are among the region's most signifying traits. Earlier in the book we noted where a duality resonates between the domestic interests of rural households and their participation in the wider public spheres of politics and economy. This duality is a necessary but sometimes disturbing condition of growth and change.[2] It also is a most lively aspect of Himalayan life.

A Grand Theory of Ecocrisis

Most conventional accounts of environmental decline in the Himalaya attribute it to the actions of the mountain villagers. According to that viewpoint, the burgeoning needs of people for fuel wood, fodder, and farmland have led to widespread forest destruction across the mountains. The loss of forests, in turn, is thought to contribute to alarming rates of slope failure, to destructive landslides, to soil erosion, and to an increase in the flow of sediments carried by the mountain rivers. Farm productivity suffers as a result of the land degradation, and many of the mountain villages are left impoverished by it. Meanwhile, downstream, the crops, livestock, and human lives in the plains are threatened by unpredictable flood waters caused in part by the denuded hill slopes. The villagers thus are blamed for eco-logical destruction that occurs not only in the highlands of the Hi-malaya but all across the plains of the Ganges River, ultimately affect-ing the lives of hundreds of millions of persons. It is a searing indictment of mountain people that has compelled far-reaching en-vironmental agendas and enormous amounts of international aid. We alluded earlier to what has come to be known as the Himalayan envi-ronmental degradation model. Because of its huge influence on the worlds' outlook on the mountains, it is worth bringing up again.

The geographers Jack Ives and Bruno Messerli wrote *The Hi-malayan Dilemma* to refute the above scenario. They summarized sev-eral decades of independent research, popular media accounts, and government policy in order to marshal support for a unified theory of Himalayan environmental degradation.[3] Then they set out to analyze their summary. Their theory described an inexorable drift from po-tential to actual instability in the Himalayan environment, from bal-ance to devastation. The two main components in their ecocrisis model were the primacy of rapid population growth in explaining the land degradation and the role of deforestation in producing the floods in the Ganges plain.

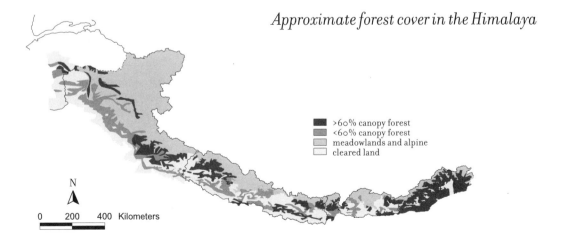

Approximate forest cover in the Himalaya

- ■ >60% canopy forest
- ▨ <60% canopy forest
- ▨ meadowlands and alpine
- ▢ cleared land

N

0   200   400 Kilometers

The prevailing characteristic of the Himalayan environmental degradation theory was its high degree of uncertainty.[4] That is understandable. No truly unified set of environmental conditions ever existed in the Himalaya. Widespread environmental problems do occur in the range, but they are never the same in different places. The causes of environmental damage, moreover, are not restricted to the subsistence villages, although they are part of it, but spill across the entire spectrum of modern society. And whereas the highland events may not be the sole cause of severe flooding in the southern Asian plains, they do seriously undermine the possibilities for human life within the mountains.

An important factor in the degradation of Himalayan landscapes, but by no means the only one, is the pressure of a growing human population on the limited resources of the range. With human numbers exceeding 50 million and with the current growth rates at over 2 percent per annum, the fragile resource base is forced each year to support ever greater demands on its productivity. Such demands in turn have led to patterns of resource use that simply cannot be sustained over the long term. In eastern Nepal, for example, where population pressures are high and where recent refugee immigrants from Bhutan have settled, as much as one-third of the farmland has been abandoned because it is rendered infertile by overuse. Many of the most fragile environments in the western regions of Garhwal and Kumaun, especially those places that host large numbers of visitors and development projects, have succumbed to the growing needs for land, forests, and water resources. Some of those places, for example the lower stretches of the Sutlej and Beas rivers, assume now an almost industrial dimension, with scarred hillsides, shantytowns, electric relay stations, godowns, and billboards.

The demand for new farmland in Sikkim and Darjeeling over the past three decades has pushed the forests there higher onto the steep mountains and have led to a suite of associated land problems.[5] All such trends predict the negative outcomes of population pressure on the land resources. Population growth in the region and its contribution to resource degradation are summarized in the theory of Himalayan environmental degradation to show how the burgeoning population occupies a central place in the spiraling pattern of social and natural tragedy in the mountains.[6] According to that scenario, the population growth leads to greater demands for forest products, which, in turn, results in massive deforestation and the loss of upland productivity. Putting the blame for environmental problems solely on population pressure is wrong, but it does have great local importance.

Poverty, on the other hand, does bring about much of the land degradation in the Himalaya. If blame for environmental problems were to be placed on a single cause, that might well be the best one. Immediate household poverty drives people to consume local resources without long-term efforts toward renewing them. The national poverty compels industrial forms of development that seek short-term economic gains based upon exploiting the natural wealth of the mountains, hence undermining sustainable conditions in the future. The regional poverty inhibits solidarity and a comprehensive approach among the Himalayan countries to deal with the region's shared environmental problems.

The size of the human population in the Himalaya continues to accelerate with concomitant poverty increases, causing greater demand for land, trees, and water. Despite the persuasions of those who view it to be negligible, population pressure has to be considered crucial to the explanation of land degradation in the Himalaya. Of course, the actual contributions of population pressure to land degradation depend greatly on where it occurs and under what social conditions. It does not have equal weight throughout the mountains. Statistics from a wide variety of government censuses, district gazetteers, and local records document the demographic history of the Himalaya during the past century. When used judiciously, they point out important geographical trends. We consulted the archival materials in all Himalayan countries and compiled them into a comprehensive overview of the Himalayan population. The numbers show a diverse and dynamic regional character to population growth.

Early Population Change

The absence of good historical records makes the determination of population size in the Himalaya prior to the end of the nineteenth century almost impossible. Consistent measurement of settlements was not generally done until the British took control of the mountains.

Nevertheless, some casual interpretations are possible based upon scattered local accounts. They show considerable settlement in the mountains as a result of the Rajput migrations into the Himalaya from the south and of dynamic population movements within the region, especially during the early expansions of the medieval kingdoms. Although the mountain villages no doubt exhibited high fertility rates, those were offset by high death rates, so only very slow population growth took place overall during the early historical and medieval periods. In some cases, cultural factors also kept the Himalayan population from growing very fast. The practice of polyandry among the Buddhist highlanders, whereby a woman had several husbands—oftentimes brothers—limited household expansion and kept intact the land a family owned. But those practices were restricted to the Tibetan cultures living north of the Great Himalaya. Elsewhere, high mortality rates from disease and poor health practices kept the population size in check.

The extraordinary shaping of the land in the middle mountains in order to produce more food, especially the elaborate hillside terraces and the formal irrigation schemes, suggests that some population pressure must have occurred early on, in tandem with the tax requirements of the princely states. But the locations of population problems are difficult to discern. Some new crops, such as the potato, were introduced in the nineteenth century. Potatoes, native to the Andes, sustained Himalayan villages located in the higher, colder areas where they grew successfully on the recently colonized marginal lands. Migration augmented the on-farm adaptations by enabling people to move in response to food shortages and to opportunities elsewhere. Apart from very general trends, though, the medieval demography is not clear.

Beginning in the latter part of the nineteenth century, though, it becomes possible to more accurately assess the major shifts in the region's population. The first real census in 1890 reports a Himalayan population of about 17 million persons.[7] That figure is conservative, though, because it does not include the populations of Bhutan or the far eastern Himalayan regions north of the Brahmaputra River. There are no reliable population data for those places, even when measured against the rather loose standards of the day. The measurements of Nepal also are at best estimations. But the figures, when applied against modern times, show a much smaller population residing in the mountains.

By the middle of the twentieth century, the number of persons living in the Himalaya exceeded 25 million. The slow but steady increase between 1890 and 1950 was not distributed evenly across the mountain region. The establishment of colonial hill stations in such places as Shimla, Nainital, and Almora spurred growth there and in the adjoining areas. The establishment of tea plantations in Darjeeling and fruit orchards in the Kulu Valley attracted migrants and additional settlement in those places. The expansion of roads and rural service centers produced growth corridors that continue to influence population movements today. Finally, the rise of urbanism, trade, and

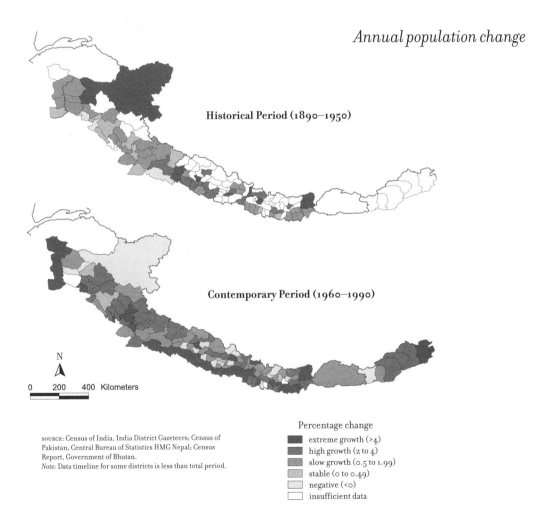

Annual population change

Historical Period (1890–1950)

Contemporary Period (1960–1990)

N

0   200   400   Kilometers

SOURCE: Census of India, India District Gazeteers; Census of
Pakistan; Central Bureau of Statistics HMG Nepal; Census
Report, Government of Bhutan.
Note: Data timeline for some districts is less than total period.

Percentage change
■ extreme growth (>4)
■ high growth (2 to 4)
■ slow growth (0.5 to 1.99)
▨ stable (0 to 0.49)
▨ negative (<0)
☐ insufficient data

A CHAIN OF EXPLANATION   137

industrial development during the later historical period forged new regional patterns of population change.

The growth rates recorded in the historical period ranged from negative values in the western Himalayan districts of Jammu and Nainital, where population declines were reported as a result of out-migration, to increases of over 5 percent per annum in such places as Srinigar and Anantang in the western and Sikkim in the eastern part of the range. A wide degree of variability in population change occupies the vast middle ground, which limits any general statement about demographic change across the mountains during the early period. The geographical patterns show, however, that the pressures of population on the land were a highly localized phenomenon.

Recent Population Change

Much of the case for an impending ecological collapse in the Himalaya rests upon the notion that the high rates of population growth during the past four decades have upset the carrying capacity of the natural environment. The burgeoning Himalayan settlements, characterized as a "population explosion," results from high fertility rates in the villages coupled with the post-1950s advances in health care and medicine. The eradication of life-threatening diseases such as smallpox and malaria, as well as preventive measures against early childhood mortality, resulted in significantly lowered death rates. Those have not been met, though, with declines in birth rates. It remains advantageous even today for mountain households to raise large numbers of children to work the land, to increase household productivity, and to care for the aged and infirm family members. A result of agrarian circumstances, expanding populations place additional demands on the fragile mountain habitats.

In the four decades since the 1950 census, more than 25 million persons were added to the Himalayan range, bringing the total number of residents there to more than 50 million people. That is a three-fold increase in the last one hundred years. The rates of growth, doubling times, and settlement densities all vary widely from place to place, though, and in accordance with environmental capacities, so a generalized Himalayan problem is less certain than are the local ones. On a world scale, the rates of population growth in the Himalaya are not unique, but given the fragility of the natural environment, the actual consequences may be quite serious.

The population of Nepal, which reports some of the most severe population problems in the range, had grown from 5.5 million in 1930 to well over 18 million in 1991. The greatest increases are reported in the outer foothills and piedmont plain, where the annual growth rates in many districts commonly exceeded 10 percent. The Kanchanpur district, located in the far western tarai, reported in 1991 the amazing growth rate of 42 percent per year. Much of that was due to migration from the hills and from neighboring India. Nepal today is the most densely settled country in the Himalaya, and one where environmental damage is thought to be the most severe. In the eastern mountains outside Nepal, similar high rates of growth are reported, almost 5 percent per annum in the case of Sikkim. In the western sector, meanwhile, three out of the four Pakistan districts lying within the Indus bend report annual growth rates of more than 10 percent.

Such extreme conditions of growth are not indicative, however, of trends reported elsewhere in the mountains. During the period 1950–90, nineteen Himalayan districts actually reported a population loss. Most of those areas are located in the middle mountains of Nepal, where people have left in large numbers to look for work or land elsewhere in the country or abroad. The variability in population change across the mountain range, as reported in the official census figures, shows a much more dynamic character of Himalayan demography than what is portrayed in the environmental literature, including in the popular theory of Himalayan environmental degradation. These accounts would have the entire mountain range overpopulated. But apart from the shared characteristics of the southern foothills, there are no truly outstanding regional trends.

The most significant population changes are tied, instead, to the geographic attributes of particular places. There is no general condition of population pressure on the mountain resources, although it is a critical factor at numerous specific localities. In order to understand the actual circumstances of population pressure, it is necessary to look at local environmental histories, politics, and the evolving conditions of the land itself. Morever, the main reasons for land degradation in the mountains, apart from natural forces, is the poverty and risk brought about by large populations, societal in-

The many faces of school children in Sikkim highlight the issue of population growth in the Himalaya. Photo by P. P. Karan

equities, and commercial developments. We have seen earlier where the history of resource appropriation by outside political interests, beginning with the feudal arrangements of the medieval period and extending through the colonial and postcolonial eras, may account for much of the environmental damage in the Himalaya. Only in specific circumstances can the occurrence of land degradation be linked exclusively to recent population pressure. Population growth, moreover, has occurred alongside other, modernization trends, in such a way that it may not be possible, in the end, to fully distinguish the primary causes of land degradation. It is most likely the result of a combination of factors.

A suite of social and natural events, acting together within a historical setting, have produced the conditions of fragility and vulnerability that describe the mountains at risk. Those factors include the size of human settlements and the aspirations of society, the occurrence of natural events ranging from tectonic activity to climatic hazards, and the securing of resource entitlements, including those which obtain over land and forests. The extent to which any one of those factors dominates the play of forces leading to environmental change is determined largely by local conditions of nature and society. The theory of Himalayan environmental degradation, in the end, described a situation that was never the case. There does exist, though, numerous local circumstances which give rise to conditions of environmental instability. They are scattered all across the Himalayan range, but conform to local not regional specifications.

Mountain-Plain Linkages?

One of the most contentious parts of the Himalayan theory of environmental degradation is the skepticism it expresses relative to the proposed role of deforestation in the occurrence of river floods in India, Pakistan, and Bangladesh. The plumes of silt that extend for several hundred kilometers into the Bay of Bengal during the flood season, observed from Landsat imagery and printed widely in the popular media, were held to be evidence of the massive sediment loads brought down by the Himalayan rivers. The river siltation was linked to the decline of forest cover in the mountains. Erik Eckholm maintained that the loss of topsoil resulting from deforestation is Nepal's number one export to Bangladesh, but one for which it receives no compensation.[8]

The environmental policies of the countries in highland southern Asia include important programs to minimize the downstream flood damage by planting trees in the mountains. *The Himalayan Dilemma* makes the linkage between upland forest change and lowland floods a pivotal point in the authors' ecocrisis model. The authors discount the relationship between trees in the mountains and floods in the lowlands, despite studies by India's Water Research Institute that show how the linkage between forest removal and increased sedimentation in the rivers is, indeed, a strong component of landscape change in the Himalaya-Ganges region.

Western scientists first began questioning the validity of linking upland forest change to lowland floods as early as the mid-1980s. Some found the answer to be contrary to popular perception or government policies. For example, the geologist Brian Carson wrote in 1985 that "flooding and sedimentation problems in India and Bangladesh are a result of the geomorphic character of the river and man's attempt to contain the rivers." He concluded that deforestation plays a minor role in downstream flood events. Two years later, the forester Larry Hamilton wrote, "Some people have blamed 'deforestation' for the catastrophic flooding . . . The real cause was too much rain." More recently, in 1995, the geographers Bruno Messerli and Thomas Hofer wrote, "[F]orest removal in the mountains plays a relatively modest role in major monsoonal floods in the lower Ganges plain."[9]

The emerging consensus among scientists is that the increase in flood damage below the mountains is due to the fact that more people reside in the flood-prone areas and not to a heightened incidence of flood events triggered by Himalayan forest loss. Hydrologists believe that many alternative land covers, such as pastures or grasslands, play just as effective a role in maintaining slope stability as do trees. The incidence of flooding in the Ganges River basin has always been high and irregular owing to the monsoon precipitation that falls among the lower Himalayan foothills. The link between highlands and lowlands continues in any case to be a recurring theme in the Himalayan environmental literature. Even if it proves to be of minor importance for the regional hydrology, the loss of forests poses immediate hardships for people eking out subsistence livelihoods in the mountains. It reduces the availability of fuel and fodder resources, diminishes the flow of freshwater springs, and curtails farm productivity as a result of the loss of organic material.

These questions about forests and floods call for some rethinking of environmental policy among the southern Asian countries. The main thrust of the government programs during the past several decades was to plant trees in the mountains in order to thwart the floods in the plains. In that case, the concerns of the lowlanders matched those of the uplanders. Much of the impetus for the conservation programs has been the understanding that such efforts were clearly in the best interests of the plains society. In fact, if the two elements are not linked, then the overall intention behind tree planting in the Himalaya may be misplaced. That is of concern for some environmentalists, who fear that without a broad base of support, the much-needed reforestation programs may lose their political backing. The pendulum will have swung against them.

Recently, the Centre for Science and Environment in New Delhi came out in support of the contention that India's floods have little to

Villagers harvesting rice in the middle mountains of Nepal. Photo by P. P. Karan

do with Himalayan forests.[10] Without a great deal of scientific evidence of their own, they tackled an environmental assumption that has attracted millions of dollars of foreign aid to southern Asia over the past few decades and risked the ire of Indian environmentalists who see the highland-lowland linkage, rightly or wrongly, as an important ally in their struggles to publicize the ecological damage in the Himalaya and to make the central governments fiscally and morally accountable for environmental restoration in the mountains. Clearly, the competing agendas in the Himalayan ecocrisis are large and diverse.

Culprits or Victims?

According to the ecocrisis model, the loss of Himalayan forestland is a result of the subsistence activities of the mountain villagers. People clear forests for farmland, cut trees for timber or fuel wood, and use plant materials for livestock fodder. Food and medicinal herbs are gathered in the forests. Livestock graze in the forest understories. As noted earlier, it is commonly held that with the growing populations, the subsistence demands on forest resources will exceed in many places the regenerative capacity of the forests. That may result in the outright clearing of forest land as well as in the more widespread "rollback" of canopy forests, where the trees are thinned and the forests degrade into less productive habitats. The results in both cases are natural and human ecosystems that may be alarmingly out of balance.[11]

In such portrayals of Himalayan life, it is easy to see the villagers as somehow the culprits in the region's environmental dilemmas. The casting of mountain farmers as ignorant and shortsighted is common in the early environmental literature. Only recently have such notions come under closer scrutiny. It is apparent that we know less about the impacts of indigenous people on forests than what was originally thought. For example, the cutting of trees for fuel wood is commonly held to be a primary cause in forest clearing. But studies show that the rates of reported fuel-wood use are extremely variable, by a factor of sixty-seven according to some studies, and overall, the contribution of fuel-wood cutting to deforestation is less important than other factors such as land clearing for subsistence farming needs or for commercial developments.[12]

The most recent environmental appraisals in the Himalaya show that in some places the negative trends are reversing, that villagers have replanted extensive areas in forest, have curtailed the open grazing of livestock, and have put considerable effort into reclaiming badly damaged land for agriculture. Moreover, many of the traditional uses of the land, such as terraced farming, rotational pastures, and common lands, are actually highly accomplished forms of environmental management that have a strong conservation background. Such efforts are now being viewed as remarkable achievements in light of the difficult nature of the mountain environment. The new assessments of land and life in the Himalaya also show that great possibilities exist for village-based environmental management by building on the traditional ways. In so doing, they highlight the fact that the villagers are not the culprits in the proposed Himalaya ecocrisis. They are, in fact, the region's greatest ally.

Finally, the theories about land degradation fall short, in the end, of really explaining the conditions of change that now characterize the Himalayan worlds. Such worlds remain, after all, highly specific places. Whereas some localities will indeed encompass many of the components of the proposed "vicious cycle" of environmental decline, others do not show strong linkages to it at all. The degree to which human society causes land degradation varies so widely across the Himalaya that a regional explanation of environmental conditions may well be impossible.[13]

Instead, life in the Himalaya reflects the highly diverse nature of land and people, the active geology, and the distinctions between the domestic needs of the hearth and the possibilities of the outside world. Within these divisions exist the uncertainties of life. Himalayan farmers are not the culprits in any broad sense of environmental collapse. Rather, they, too, are its victims. They are held hostage by the conditions of poverty and power that describe the overwhelming demands of population, and now of industrial expansion, in the Himalaya. In order to more fully comprehend the current dimensions of environmental change in the region, it is necessary first to understand the ecological fragility of so much of the region. It contributes to the levels of human impoverishment, which in turn may help to explain why the Himalayan mountains are at risk.

Commercial vehicles negotiate a landslide on a western Himalayan road. Every monsoon season, large sections of mountain roads are washed out, creating hazardous driving conditions and financial costs for Himalayan countries. Photo by D. Zurick

Fragility and Instability

THE COMMON THINKING equates forest loss in the Himalaya with an overall deterioration of the mountain environment. That viewpoint, though, is too simplistic; in fact, a number of land dynamics are involved. But it does focus our attention on the interlinked problems of resource scarcity, soil erosion, and the loss of biological productivity in the uplands. These elements constitute significant modifications of the mountain worlds—where forest edges recede farther upslope, gullies scar the hillsides, and farmland is abandoned. Such disturbances in the landscape are now apparent to even the most casual visitor in the region. Of course, they are the very nexus of life for many native Himalayan people.

The dramatic exhibition of many of these forces in such occurrences as the landslides that bury entire villages and gouge large chunks out of the mountainside makes a compelling testimony to the

9

stressful conditions in the Himalaya. Other indicators of the mountains' fragility, less evident perhaps in the landscape, may be even more devastating for the mountain communities. The decline in food production, for example, is a farming concern across the range and is a major cause of poverty. It may not be readily seen in the landscape, but it is undeniable in the limited diets of many Himalayan people. Under the traditional, organic systems of mountain agriculture, in which forests and farms are joined under carefully managed systems of composting and recycling, the loss of farm productivity is a clear indicator of damage not only to the forests but to entire mountain ecosystems.

The most pronounced landscape shifts produce burdensome conditions for life in the mountains. For example, where forests are degraded the villagers often must walk several hours daily in order to procure sufficient fuel wood or livestock fodder. The image of a Himalayan peasant carrying a huge backload of plant material along a steep mountain trail has become a lasting icon for the region. In a naive way, the image serves as a symbol of human strength and resiliency; more properly, though, it is a sign of human and environmental poverty. Village women must wait patiently for water at the village springs. It makes a great photograph, but on some days it may take hours to fill a single vessel. The women attribute the delays to the shrinking of their woodlands.

Such dire circumstances are brought on not only by human activities but also by the dynamic natural conditions that distinguish the Himalayan environment. The notion of a simple progression from stable conditions to vulnerability, as proposed in the Himalayan ecocrisis model, is misleading. There exist no real conditions of stability anywhere in the mountains. The Himalaya, in fact, are one of the most dynamic places on earth. The high rates of tectonic activity, the inherent ecological fragility of the natural habitats, the vagaries of climate, and the capricious conduct of human society all serve to ensure that the mountain landscape will not remain immutable.

The issue, then, is not with change, which is necessary and unavoidable, but with its general direction and its long-term consequences for the native people and the natural habitats of the Himalaya. The transformations that continually shape the landscapes are most confounding because they are so widely divergent, and they remain largely unpredictable from one place to another. But the simple conversion of forest to farmland, a key element in the Himalayan ecocrisis model, need not lead to land degradation at all. Rather, it may reflect the attentive and sustainable use of the landscape by the villagers.

Tethys Rising

Several million years of steady grinding movement were required for the mountains of the Himalayan range to attain their present stature from their geological roots in the Tethys Sea formations. The fact that the mountains continue to grow, despite high rates of weathering, highlights the fundamental idea that the Himalaya conforms to an order of progression that is well beyond the reach of human experience. The mountains lie in a collision zone between the northward drifting tectonic plate of southern Asia and the inner Asia plate. It is one of the most powerful seismic areas in the world. As a result of its northward drift, India is moving against Tibet at the rate of about five centimeters a year. That movement results in as much as ten centimeters of mountain uplift in some places each year. But the rise of the mountains is not without its perilous side. The same forces that cause the mountains to form also create a highly unstable geological environment that contributes to a hazardous landscape filled with risks for human settlements.[1]

The geological maps of the Himalaya that commonly appear in books and atlases show a confusing maze of multidirectional arrows and arcs. The lines on the maps point to the locations and trends of the numerous thrust faults, fractures, and rupture zones that rip through the mountain landscape. These geological maps look like hieroglyphics. They show a complex alpine system filled with tremendous stress and motion. The maps fix with clustered icons the locations of the violent earthquake epicenters, where each year tremors rock the villages, causing massive landslides, damaging settlements, and killing people. Between 1961 and 1991, scores of recorded earthquakes claimed the lives of hundreds of mountain people. In some cases, entire villages were buried beneath avalanches triggered by the quakes. Elsewhere, lonely shepherd camps succumbed to falling rocks, or porters walking along the mountain precipices disappeared as the ground beneath them literally gave away. But only the most severe episodes make the international news.

Every so often, though, an earthquake of truly global proportions rocks the Himalayan region. The most recent of these occurred in Assam in 1897, in Kangra in 1905, in Bihar in 1934, and again in Assam in 1950. All these earthquakes recorded magnitudes of greater than 8.4 on the Richter scale and were reported around the world. The active earthquake zones, where major slippages of earth have occurred along the fault lines, extend over a 500-kilometer area. Some places, though, have escaped recent major earthquakes. They represent seismic gaps in the alpine landscape. Considerable tectonic pressure has built up in such places, making the future event of a huge tremor all the more likely there.

The frequency with which earthquakes and related catastrophes occur in the Himalaya imperils human society and its various constructions. In addition to the damage they do to homes and farms, the earthquakes block roads and other infrastructures, and call into question the placement of the large dams built along the active fault lines.[2] The entire region is described by geologists as a "high energy environment," which means that in addition to the notorious seismic events, a great deal of earth material is transported down the mountainsides by the quieter movements of weathering and erosion. The slide of the mountain slopes is due to their length and steepness and to the erosive power of rain and wind. Most notable of the materials that are eroded and then carried away by rivers is topsoil.

The geology of the Himalayan region predicts that natural rates of soil erosion will be high regardless of the level of human interference. The sizable earth tremors cause extensive sections of slope land to slip away, resulting in large debris flows. The catastrophic bursting of highland lakes, dammed by unstable glacial moraines, may cause the quick release of trapped water in the high mountains. The deluge of water and sediments that result from these burst lakes travels onto the lower rivers and valleys, possibly causing devastating floods.

One of the most notable glacial lake outbursts occurred in the Annapurna Mountains about 600 years ago. It sent five and a half cubic kilometers of earth plunging into the Pokhara Valley in central Nepal.[3] The magnitude of that catastrophe is hard to imagine, but it is a recorded event in the central Nepalese landscape. More modest examples of such glacier outbursts occur in a more regular fashion all across the high Himalayan range. In Nepal's Khumbu region, there have been as many as five outbursts in living memory. The most recent occurrence, in the early 1990s, sent torrential floodwaters down the Dudh Kosi River, destroying homes and trails and leading people in Kathmandu to question what might have happened had it occurred during the height of the trekking season when hundreds of foreigners were on the trails. The damage to the country's tourism image was worrisome enough for some officials to ask for new studies of the problem. Meanwhile, remote villagers live day to day under such unpredictable and potentially tragic circumstances.

At a closer scale, where the hill slopes are steep and the rainfall is especially heavy, for example in the eastern sectors of the range, the natural potential for soil erosion is quite great. During the monsoon season, the rivers of the Himalaya run brown under heavy sediment loads. Much of this coloration comes from the loss of topsoil, which is removed from some upland farms at an annual rate of one or two millimeters. Since the natural weathering processes in the mountains produce much less soil than that each year, future management of the farms is brought into question. Meanwhile, the farmers constantly work to maintain the fragile soil base of their fields. The construction of terraces on the steep hill slopes, a hard job that requires constant maintenance, is a long-standing adaptation of mountain cultures to the ever-present problem of soil erosion.

The great majority of the landslides that occur in the Himalaya take place on hillsides over 30° steep. The most spectacular ones are triggered by excessive rainfall. The Darjeeling landslide in 1968, for example, which buried hundreds of villagers, occurred when more than half a meter of rain fell in the area during the course of a single day. Immediately after the monsoon season, which takes place during June to September, the landslide scars are most evident on the hillsides. In some cases they result from poor land management on steep slopes. But landslides also are natural events in the mountains and will occur without a human presence. When they take lives, the landslides are tragic, but they are not necessarily a product of the misuse of the land by human society.

The theory of Himalayan environmental degradation does, however, consider soil loss in the Himalaya to be a problem of mainly societal origin. When forests are depleted and pastures overgrazed, the protective vegetation cover is ruined, exposing the topsoil to water and wind damage. Traditional cultural practices clearly may produce the shifts from canopy forests to grasslands. But roadcuts also undermine the slopes and cause landslides along their corridors. They add to the risky aspect of mountain life, but stem from economic policies beyond village control. At the same time, farmers carefully manage their grazing lands, protect their forest groves, and shape the tilled slopes into terraces, which, under good maintainance, prevent soil loss. The human time frame of such efforts is nothing, though, compared to the geological progression of tectonic forces, which slowly uplift the mountains, and of natural weathering, which wears them down. Change, rather than equilibrium, is the premier feature of the Himalayan landscapes. But it may happen without notice.

In Nepal, where the problem of watershed degradation is widespread, the country loses an estimated 240 million cubic meters of soil annually.[4] Much of that ends up in the tarai plains, where the riverbeds are rising at a rate of fifteen to thirty centimeters annually. The filling of the streams with sediments causes them to overflow their banks. The monsoon floods in 1993 were especially devastating, killing more than 1,000 people and 25,000 livestock, and inflicting a damage of more than $75 million U.S. Notwithstanding the predictions of earth scientists who claim that such events are natural and inevitable, the scenes of floating animal carcasses, buried homes, and loss of human life evoke the tragic dimensions of land degradation and floods. The proposed solution in Nepal is to repair watershed damage by reforestation in the hills. Yet, by the early 1990s, only 5 percent of the deforested areas had new tree plantings. Meanwhile, and perhaps ironically, the fertility of the tarai soils is actually enhanced by the annual accretions of sediment. The migration trends in Nepal also show people moving from the hills to the lowlands, as if in response to the actual movement of topsoil.

In the Khumbu region of northeastern Nepal, famous for being the site of Mount Everest and the home of the Sherpa people, a large number of scientific and media reports have described the degraded environmental conditions in the mountains. Yet recent studies by geographers and other scientists in Khumbu show that soil erosion there is quite minimal and is linked mainly to natural rather than to cultural factors.[5] The landscapes of Khumbu, and other places like it, have been greatly modified by human settlements but not necessarily ruined by them. In looking for forests among the peaks and high pastures, the Himalayan visitors may seek a false wilderness. The mountains have been inhabited for a very long time. Their landscapes reflect the aspirations of people as well as the processes of nature. They are lived-in places, not wild ones, managed by native people who are astute environmental observers.[6] Viewed in such a way, the transformed landscapes of the Himalaya may be seen as places which serve to sustain human life, not to prohibit it. Only in those places where human activity seriously surpasses the carrying capacity of nature do we find evidence of truly degraded environments.

In some places, those circumstances are linked to subsistence pressures in the villages. Other activities, though, such as constructing roads, dams, or power lines, and the commercial felling of forests, also produce unstable conditions that may undermine the stability of mountain slopes and accelerate the rates of soil loss. The damage is most likely where the hillsides are steep and the commercial projects are poorly designed. In such cases, the destructiveness of infrastructure developments may greatly outweigh that of subsistence activities. For example, environmental studies in the eastern region of Sikkim and in the mountains of western Garhwal both show that on average every single kilometer of constructed roadway causes at least two major landslides. Road building in Nepal, meanwhile, produces up to 9,000 cubic meters of landslide per kilometer. Overall, it is estimated that a single kilometer of road in the Himalaya will eventually cause 1,000 tons of land to be lost because of slope failures.[7]

The singular experience of traveling on the Himalayan roads, especially during the monsoon months, shows that such statistics accurately describe roadside conditions. From a distance, the roadways snake along the mountain contours and cause a visible line of gully scars and debris outfalls. Up close, washouts and land slips frequently block the way ahead. Such conditions attest to the unstable conditions produced by the roads. It is estimated that up to one-fourth of all river valley roads in the Himalaya are washed out every four to five years. The environmental implications of that are quite serious. Moreover, the fact that the roads need continual repairs creates serious financial problems for the mountain countries.

While it may be difficult to separate out the exact causes of soil loss, it is apparent that the large infrastructures such as road alignments and the other commercial innovations do more to destabilize slopes in many localities than do the foraging or farming efforts of villagers. And in the sum of the land-degradation equation, it also is clear that natural events such as earthquakes, landslides, glacier-lake outbursts, and other seismic events produce high rates of slope instability and soil erosion. The monsoon climate also exacerbates soil erosion, by dumping heavy precipitation on the land all at once.[8] Additional climatic factors such as strong winds, hail, and frost add to the hazards of the mountains. The diverse topography leads to abrupt changes in local climate, which adds to the severity of mountain hazards.[9] The scale of such natural events may simply overwhelm in many places the contributions of humankind to land degradation. Nonetheless, the human communities must continually cope with these happenings as well, which makes life there all the more precarious.

A Valley in Sikkim

The uncertain environments that confront communities across the Himalayan range take many forms, but common to all is the close association that exists between society and nature. A small valley in eastern Sikkim, one of a series of study sites, shows how that relationship enters into the environmental debate.

The Richu Khola River is located in the middle mountains of Sikkim, about ten kilometers from Pakyong, a small market center serving the local population. The watershed covers an area of 1,581 hectares, at elevations varying from 625 to 2,591 meters. The area is representative of an extensive region from eastern Nepal to western Bhutan. The watershed lies in a rough north-south direction on the southern aspect of the main ridge line.

The area is encircled by greatly dissected longitudinal ranges. The drainage pattern is dendritic, and the tributaries of Richu Khola are mainly seasonal streams. Richu Khola itself is perennial. The northwestern and northern parts of the watershed are at altitudes above 1,900 meters; the southern area is about 625 meters near the junction of Richu Khola and Rangpo. The general topography of the watershed is highly undulating, with slopes ranging from 10 to 15 percent to very steep, 50 to 100 percent. There is a conspicuous absence of land that falls within the slope range of 1 to 10 percent, which is the ideal range for intensive agriculture without steep terraces. That is mainly due to the abrupt rise of mountains and ridges from the drainage channels that flow through deep gorges. The amount of rainfall in the watershed is very high, annually averaging about 140 inches. Maximum rainfall is received between May and September, with a peak period in July. Richu Khola flows at a high velocity; it is flashy in character and carries much silt and debris. The small streams, meanwhile, pose large problems by transporting sediments and by severely eroding their sides, thus causing the loss of a lateral support to the hillsides. Owing to stress, cracks develop in the earth in the upper reaches.

Agriculture has become increasingly important in the middle and lower zones of the Richu Khola watershed in the past fifty years. Most of the natural vegetation was cleared during the process of reclamation for agriculture. Richu Khola watershed is a good example of a Himalayan area where human activities—expansion of agriculture and destruction of forest cover—combined with heavy precipitation and steep slopes have resulted in serious environmental problems. The watershed has less than 20 percent of its land classified as suitable for agriculture, but 43 percent was being farmed in the early 1990s. The farmers are growing crops on extensive areas of land that, for reasons of soil makeup or slope steepness, should not be farmed. Moreover, while the watershed has three-quarters of its land suitable for permanent vegetation, at present only about half is under trees. The forest area has decreased during the past fifty years. Sikkim once enjoyed luxuriant vegetation of different types. Human interference has led to considerable adverse effects on the density and distribution of natural forests.

Farming is the mainstay of the 1,979 people living in the watershed. The area under agriculture has increased over the last five decades, and at present agricultural holdings vary from two and a half to eight hectares per household. That is quite high compared to other Himalayan regions. Most of the agricultural land is rainfed. The main crops grown in the watershed are rice, maize, wheat, and a local grain called *marua*. Mixed farming at a small scale is practiced by farmers as an informal risk-minimizing strategy to ensure a minimum level of

A QUESTION OF BALANCE

Rice planting in the irrigated fields of Sikkim. Photo by P. P. Karan

production irrespective of the weather conditions. But land management practices in Richu Khola are generally not adequate to meet the environmental challenges. For example, terraces are laid out but they are not well designed and maintained. The terrace risers are very steep, which at times damage the bench terraces. Shoulder bunds are provided, but water is disposed from terrace to terrace, thus causing damage to the risers. Fodder crops are grown near the terraces and drainage channels, and they provide a limited buffer zone of vegetation.

More than one-half of the area cleared for farming is not really suitable for agriculture. The slopes are too steep and the soils are thin. Crop yields are low, and the land is subject to severe erosion. Horticulture crops are not grown in the watershed, although there is adequate scope for their development. Likewise, farmers have not done much to develop pastures and improve the quality of livestock.

Considering the present land-use patterns, land capability surveys, and natural processes in the watershed, one can say that it is possible to develop an integrated land-use plan for environmental management and development of the Richu Khola watershed. That plan could be viewed as an attempt to bring overall stability to the fragile ecosystem as well as to enhance the production system. The proposed plan would put emphasis on (1) increasing the intensity of cropping to increase production, since there is no scope for expanding the area under cultivation; (2) bringing land not suitable for growing rice under horticulture crops such as cardamom or citrus; (3) developing community pasture based on land-capability classification and animal husbandry; and (4) developing farm forestry to protect the steeper slopes. Under such a plan, the landscape would best accommodate a forest cover on 60 percent of the area and agriculture on much of the remainder. (The most serious landslide areas likely cannot be reclaimed in the near future.)

Farmers can be encouraged to grow horticulture crops on the steeper land. To avoid hardships to the farmers by introducing horticulture, which normally takes about five to seven years to yield fruits, intercropping of the terraced or trenched areas could be encouraged. Cash crops such as vegetables, potatoes, root crops, and strawberries could be cultivated. This would increase cropping intensity and ensure a regular income until the main horticulture crop began giving a return. Depending on altitude, citrus, mango, guava, peaches, plums, pears, and walnut may be suitable for small orchards within the watershed. Tropical and subtropical species may be better grown on the south- and southwest-facing slopes, while temperate fruits may be more suitable for higher elevation and north-facing slopes. The great environmental diversity of even such a small place as the Richu Khola watershed gives the villagers a wide range of potential livelihood options. That is one of the most compelling features of mountain communities all across the Himalaya. Their choices actually may be varied and profitable once environmental conditions are stabilized.

Mountains at Risk

As we have said, the geology and climate of the Himalayan range combine with the impacts of human settlement to produce dynamic environmental conditions. The basic life-support systems of soil, water, and biota are the limiting factors in the mountain ecosystems. Their scarcity or degradation forces mountain communities to live frugally. The soil is fragile because of the steep slopes, the high rates of runoff, and the rapid weathering. Water is fragile because of the variability of precipitation, the growing scarcity of drinking water, and the lack of reliable water-storage systems. The biota is fragile because of its delicate adjustments to the restrictive and ever-changing conditions of the natural habitats. Where species diversity is high and the number of native plants is great—such as in the eastern Himalaya, where almost 40 percent of the more than 9,000 biotic species are endemic—a critical issue is the potential loss of biological diversity. The entire eastern region has been designated a world-class biodiversity "hotspot" owing to the impending threats to its natural habitats.[10]

It is noteworthy that the fragile nature of the Himalaya is commonly regarded as one of its most defining characteristics. That assessment seems incongruous against the overwhelming stature of the mountains and the resiliency of the people who inhabit them. It nonetheless compels much new research and development policy on the interlinked issues of environmental preservation and cultural survival. In 1994, a scientific forum was held at the International Centre for Integrated Mountain Development in Kathmandu to consider the causes of mountain ecosystem instability.[11] The problems of environmental change, natural hazards and disasters, and unsound development practices were considered at that conference. All these problems were found to be prevalent across the entire Himalayan range.

But the specific conditions that give rise to the uncertain circumstances in the mountains are more diverse even than those. They can be organized around three main considerations.[12] First is the quality of accessibility. Owing to terrain conditions and to the lack of roads, many places in the mountains remain isolated from national or regional infrastructures. That limits the dependable supply of outside resources, causing communities to continue to focus internally on local resource systems. Where population pressure is great, the local carrying capacity may be surpassed. Conversely, enhanced accessibility provides some basic needs through the marketplace, but it also contributes to the growing commercial pressures on mountain resources. That duality causes a fundamental tension for many communities seeking to strengthen their regional ties.

The mountain communities that are dependent upon local resources for their livelihoods are compelled to diversify land uses and to work toward the continued regeneration of the natural environment. If they do not, then the low carrying capacities of the land will continue to constrain productivity and to limit social aspirations.

Marginality is another defining characteristic of the fragile Himalayan environment. Mountain habitats are marginal in the sense that their levels of productivity are curbed by such factors as altitude and slope. They also are considered marginal in the sense that they are geographically or otherwise removed from the mainstream developments of dominant societies. Occupying the spatial edge of the world, particularly where infrastructure is concerned, precludes mountain communities from participating in certain national or regional developments. It also promotes, or at least allows, a general disregard of mountain problems by decision makers residing in the lowland political centers.

Finally, the factors of environmental risk are offset somewhat by the unique combination of resiliency, diversity, and comparative advantage found in the mountains. These latter features compose the basis for cultural adaptation among the traditional societies in the Himalaya. For example, the altitudinal zonation of mountain worlds, described earlier in the book, contributes to a great medley of habitat conditions. People may reduce the risks associated with the hazardous conditions in the mountains by spreading their assets and subsistence efforts across a wide range of environment types. The result is the complex systems of economy that prevail in the Himalaya.

The mountain environments give the local economies the benefits of comparative advantage by providing environmental qualities not found in the plains. The domestic needs of the Himalayan households, oriented toward key mountain resources, incorporate this important aspect into their daily foraging, herding, and farming pursuits. The wider market economy now also views the comparative economic advantages of the Himalaya to be one of the region's outstanding assets. For example, the temperate climates of such places as the Kulu Valley, Shimla, and the district of Kinnaur in the western mountains has allowed the formation of specialized fruit-based horticulture regions, which have a large and lucrative market in the southern plains. The fast and powerful rivers of the Himalaya, harnessed by dams and tunnels, have enormous potential for hydroelectric generation. The splendid alpine scenery and the unique mountain cultures are the basis for tourism. The montane forests are rich in much-sought-after timber and medicinal herbs. Those who make use of these possibilities can exploit the unique attributes of the mountains for the purposes of commerce and economic gain.

The compensatory features of environmental diversity and comparative economic advantage in the Himalayan worlds are determined largely by the aspirations of human society. To some extent, they can offset the conditions of risk produced by geological or climatic events and human interferences. But the alarming processes of deforestation, soil erosion, water imbalances, and reduction in biodiversity also are rooted in some basic social forces that will continue to adversely affect life and land in the mountains.

The impoverishment of mountain society and the degradation of the natural environment are two infelicitous sides of the same coin. Notably, the accelerated commercial exploitation of environmental resources may well jeopardize the continuation of local cultural adaptations. In so doing, it may also undermine the conservation efforts of indigenous people. That already is widely apparent where the new institutional controls over land management, or the commercial appraisals of resources, exclude the common lands management approach of traditional communities. In such cases, both the land and the people will likely suffer.

Population growth is often cited as a major social cause of the burgeoning demands for forest resources. We have shown earlier how the increase in the numbers of persons living in the Himalaya has a long history and is not evenly distributed across the range. In the prehistorical and medieval periods, when population growth was fueled by migrations, population pressure led to new styles of agriculture and to more intensive uses of the land. During the contemporary period, when rapid population increases have occurred largely as a result of health innovations, the exploding human numbers have led in places to alarming declines in available resources and to accelerated land degradation. Under the conditions of fragility described above, these new demands make the Himalayan worlds even more vulnerable. We will look at such patterns of population pressure on the land in the next chapter.

The commercial pressures of modern development, meanwhile, add their own conflicts to land use in the mountains. The market forces developed during the colonial period, for example, treated the mountain habitats as inexhaustible sources of timber and other forest supplies. The more recent efforts to extend transportation and commercial linkages to rural places have encouraged the production of cash crops and the further extraction of such resources as timber, water power, and minerals. Where those activities are not well-monitored or regulated, the environmental impacts may be devastating. The elimination of key resources is much more likely under such circumstances, and the diversity of mountain systems may be grossly simplified by the new market economies.[13]

To overcome the constraints imposed by conditions of fragility and instability, the traditional societies of the Himalaya rely upon a great wealth of cultural knowledge. The movement of livestock herds, the social regulation of forest lands, the distribution of farm fields across numerous microenvironments, the tillage methods, and such alternative economic efforts as trade and handicrafts combine in the livelihood repertoire of indigenous people to produce a comprehensive approach to life. Nonetheless, the thresholds of many natural resource systems may already be surpassed in some places, accounting for the current conditions of serious land degradation. As we shall see in the next chapter, however, the geographical pattern of subsistence pressures on the land remains complex and highly localized.

For development purposes, the constraints imposed by the fragile and unstable mountain environments are thought to be overcome by financial subsidies and technical inputs. Biochemical additions to the soil aim to increase or maintain its productivity. Tree-planting regimes that rely solely on exotic species seek, not to replicate the native forests—which contain an abundance of species and provide diverse products—but rather to promote instead a reliable supply of harvestable timber. The harnessing of rivers by irrigation schemes, dams, or run-of-the-river diversions are meant to support industrial farming, electrify the villages, and generate energy for export. These varied efforts, all of which seek to overcome some of the most obdurate characteristics of the Himalaya, impart an essentially Western approach to the classical problem of low environmental productivity. The measure of their success rests mainly in new evaluations of sustainable development.

Meeting the basic needs of most Himalayan people, rather than the rising material aspirations of only a few, is a basic tenet of the sustainable-development perspective. Another is the reliable and regenerating use of local environmental resources. The participation of local people in the design and management of development programs is a third. According to the sustainable-development models, the proposed solutions to the problems of instability in the Himalaya depend upon the close match-up between local mountain resources and their long-term use by resident communities. This position is antithetical, however, to many of the current big development projects, which rely mainly upon external inputs and foreign assistance for their execution. The wealth that such developments may produce is largely meant to be sold to urban places rather than to be used in the organic mountain farms.

Cattle graze in a forest opening in Himachal Pradesh, India. The open grazing of livestock contributes to the degradation of Himalayan forests when appropriate land-management practices are absent. Photo by D. Zurick

The Weight of Life

THE HIGH DEMANDS on Himalayan resources stem from both the subsistence requirements of native populations and the commercial developments of the national economy. They combine with such natural factors as the geological rate of mountain uplift, seismic action, and the physical base of slope processes and erosion to produce the adverse conditions of land degradation that prevail in many places.[1] Both society and nature must therefore be considered in modern assessments of environmental change in the region.

The role of resource use and patterns of human development in the landscape, which from a cultural viewpoint are critical factors in the transformation of the land, are as diverse in the Himalaya as is the natural environment itself. Much of what we know about the impact of human activities in the mountains comes from studies conducted among communities residing in the middle-mountains zone of

10

Nepal.[2] They indicate dense human settlements and troubling environmental conditions. More recent studies among the high plateau regions in the north and in the lowland areas of the southern tarai show those places also to be damaged by trends adverse to sustainable development. In all cases, the scope and intensity of economic activities must be considered against the environmental and cultural conditions that prevail locally.[3]

Where the increasing pressure of human needs exceeds the carrying capacity of the environment, the possibilities for land degradation may be quite high. The sustainability of the mountains under such circumstances may be threatened. That is also true for the agricultural systems of the Himalaya, which in many areas already show signs of decline. The problems are linked to dwindling farmland allotments, to stagnating or decreasing yields, and to the abandonment of the traditionally diverse and organic farming methods. Commercial use of the Himalayan environment in other places has contributed to systemic changes in the natural ecosystems and to other changes in the landscapes that are now so widespread as to give the mountains a global importance.

The uniquely fragile conditions of the Himalaya apply to the primary areas of farm production, just as they do to the wildlands. Agriculture links land cover, water, and human activity within a narrow resource base that is especially vulnerable to the changing conditions of life. Moreover, the conditions of agriculture in the mountains are tightly fixed by ecological factors, corresponding to environmental zones and cultural adaptations. But overall, the mountains contain highly complex systems of food production and resource management.

Sustainability in such cases may be best considered in terms of the intensified use of resources.[4] We can think about this in regard to both the domestic subsistence economy that still prevails widely across the Himalaya, and the market economy that is fast penetrating the mountain world. The pressures on mountain resources that derive from human activity may be considered according to their origins in either or both of those economies.

Demand on Landholdings in Phalabang Village

Farming is still the most dominant feature of the Himalayan cultural landscape. It must be considered in relation to all the components of the mountainous land cover. The organic farms transfer nutrients from the forests and grazing lands to the crop areas and then recycle them in composting practices so as to sustain the harvest yields. Matching the capability of the land with the level of agricultural intensity requires a proper ratio of farmland to forest. The village of Phalabang in western Nepal, which in this important way is typical of many farms in the middle mountains, shows how livestock grazing and fodder collection in the forests are tied to regional landscape ecologies. More generally, patterns of human settlement and land use contribute to a Himalayan-wide view on farmland pressures. In combination with modernization efforts, they indicate the prospects for food security in the region. Phalabang provides one small example of this.

The landholdings in Phalabang are very small, averaging about a hectare per family. That is typical, though, of much of Nepal, where per capita agricultural holdings range from less than a tenth of a hectare in some of the western districts to only a third of a hectare in the scattered hill districts in the east.[5] The farmland in Phalabang tends to be split among several fields, which are distributed across varied environmental zones, extending from the fertile rice paddies situated along the bottomlands of the Sarda and Luwam rivers to the upland dry fields on the high slopes of the Mahabharat ridge line.

The fragmentation of farms has produced a village patchwork of land use. There is little opportunity for additional land clearing nor has there been for the past half century or so. The current organic farming methods require sufficient access to forests for the purposes of compost and nutrient cycling, but the forest resources in Phalabang were seriously diminished during the period from 1960 to 1990. Agricultural scientists recommend that five hectares of forest be available for every hectare of farmland for the transfer of green compost and livestock dung to be most efficient. In Phalabang, though, the ratio is less than two to one, well below the optimum level.

The low forest to farm ratio in Phalabang reflects the fact that over the past several decades the expansion of cultivated land has occurred at the expense of canopy forest cover. Aerial photographs of the village

The autumn grain harvest in the valley of the Sanghla, India. Photo by D. Zurick

in 1953, 1978, and 1990 show the progressive loss of forests to shrublands and pasture. In 1953, village forests composed almost one-half
of the village territory; by the end of the 1980s they occupied less
than a third of the territory. During the same period, the area covered
by badly managed pasture and wasteland increased several fold.[6]
Meanwhile, the number of village livestock has continued to increase,
far surpassing the capacity of the wildlands to produce sufficient fodder to adequately feed them.

While the overall impact of such trends on slope stability may not
be certain, it is clear that the potential loss of fodder and biomass is a
problem for farm productivity. The trends discovered in the village of
Phalabang are perhaps unique in their specific detail but they typify
processes that are replicated across the mountains. In the surrounding 10,500-square-kilometer Rapti Zone, for example, similar conditions prevail. Apart from the Dang district, which occupies a fertile
dun valley situated in the outer foothills region of the zone, all the
mountain districts of Rapti reported household landholdings close to
the size of those found in Phalabang. There exist greater amounts of

forest land in the northern part of the zone; the Rukum district, for example, has three hectares of forest land for each hectare of farmland, but in the most densely populated regions in Rapti, the agriculture land to forest land ratio is very low, similar to that found in Phalabang. Moreover, many of the intact forests in the northern districts are located at a great distance from the farm fields, more than four kilometers, making them difficult to reach for purposes of fodder collection or animal grazing.[7]

The agriculture land to forest land ratio is one measure of stress within the organic farming systems of the Himalaya. It works best to illuminate conditions at local scales, where the geographical linkages between farms and forests are well known. In Phalabang, and in many villages like it, the farmland-forest ratio is much too low for a sustainable economy. Other measures of stress in the farming systems include direct population pressure on the crop land and on forests. These pressures also show the intensity of land use in Himalayan subsistence economies. Such measurements were determined at the district level for the entire Himalayan range, utilizing available census and archival information. They show the broad trends that have shaped the mountains into a heterogeneous and highly dynamic landscape, where environmental pressures are great in some places but relatively minor in others.

Pressure of Population on Himalayan Farms

In order to assess the quality of agricultural conditions in the Himalaya, it is necessary to consider such factors as soil fertility, water, crop type, labor, and other inputs as well as the pressure of human numbers on the limited available farmland. These variables set the factors of production and will vary widely across environmental zones and cultural territories. Efforts commonly are made to assess them and to ascertain local agricultural capabilities—ultimately to modify the farming systems in order to make them more productive.[8] The measure of population pressure on crop land is a key variable in considering the rates of land-use intensity and is a function of both population change and shifts in the amount of reported farmland.

Archival records show that during the historical period 1890–1950, a steady increase in crop area occurred throughout the western

Himalaya. This was most pronounced in an east-west belt along the southern limit of the middle-mountains zone and among the districts that occupy the western outer foothills region. Hence, we find in such places as the Dehra Dun, Garhwal, and Almora districts some very high rates of farmland expansion during the historical period. The area under crops in these places increased at an average rate of more than 2.5 percent each year. In some instances, such as in Garwhal, the historical rates of farmland expansion exceeded 10 percent a year.

In six isolated western districts, however, there occurred a slight decline in the area under cultivation. That is attributed to a variety of factors, which include declining soil productivity and other forms of land degradation, outmigration for employment and consequent land abandonment, increasing levels of urbanization and infrastructure development, war and political unrest—such as we find in Srinigar and the neighboring Baramulla districts, and to the emergence of off-farm opportunities. In some of the western Himalayan districts, such as in Shimla, the loss of crop area is linked also to the development of horticulture, especially apple orchards, which have supplanted in many places the food-grain lands. Elsewhere in the Himalaya, in Nepal and in the eastern regions for example, insufficient historical data exists to reconstruct the historical agricultural landscape. But clearly in those places, owing to the developing geopolitics discussed earlier in this book, a steady expansion of farmland has occurred across the middle-mountains zone.

A more complete picture of farmland change in the Himalaya emerges from the information covering the contemporary period, 1960–90. These data, when mapped, show a highly complex mosaic of regions where crop-land gains and losses have occurred. A few regional clusters appear in the maps. High rates of crop-area loss are reported throughout the Himalaya, but are most concentrated among the mountain districts in the western region. With the exceptions of the Pithoragarh district along the western Nepal border, where a slight increase in crop area is reported, and Tehri-Garhwal, which exhibits stable conditions, all of the mountain districts in the Kumaun-Garhwal region show significant to high rates of farmland loss. They are alarming trends that go against the intuition of the ecocrisis model that has farmland expanding all over the Himalaya.

Major areas of crop land have been put out of production in parts of the western region because of failing land productivity. The exhaustion of soil fertility due to overexploitation and the inclusion of marginal lands of low fertility in the cultivated areas have led in such places to serious reductions in crop yields over the past decades. In the mountains of Kumaun, for example, yields of maize, millet, and oilseeds have dropped to very low levels.[9] In light of the declining trends across the western Himalaya region, the Indian and Pakistan governments have targeted farming systems improvement as one of the major development initiatives in the mountain region. Since little new land is available, most of the agricultural programs are aimed at increasing the intensity of land use through intercropping, new crop

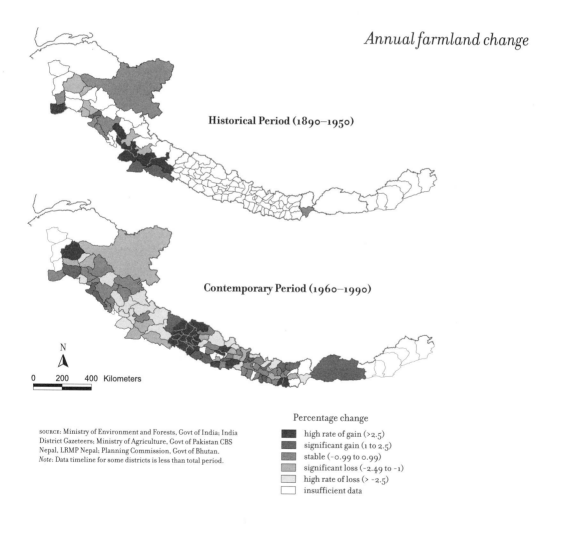

Annual farmland change

Historical Period (1890–1950)

Contemporary Period (1960–1990)

N

0   200   400 Kilometers

SOURCE: Ministry of Environment and Forests, Govt of India; India District Gazeteers; Ministry of Agriculture, Govt of Pakistan CBS Nepal, LRMP Nepal; Planning Commission, Govt of Bhutan.
Note: Data timeline for some districts is less than total period.

Percentage change

high rate of gain (>2.5)
significant gain (1 to 2.5)
stable (-0.99 to 0.99)
significant loss (-2.49 to -1)
high rate of loss (> -2.5)
insufficient data

Landscape changes result from farmland expansion in a forested area of Arunachal Pradesh. Photo by P. P. Karan

types, and the application of technologies such as farm machinery, fertilizers, and pesticides.

Across the border from India, in the western mountains of Nepal, a cluster of fourteen districts covering an area of more than 25,000 square kilometers exhibits very high increases in farmland. This historically isolated part of Nepal exhibits some of the lowest development indicators in the country. The region also shows low levels of population density, and less than 10 percent overall of the area under cultivation.[10] With such low rates of agriculture, even minor increases in crop area may generate significant percentage increases. Significant real increases in farmland have occurred, however, as a result of several factors. The region is targeted for major infrastructure developments, including roadways under construction to Darchula, Baitadi, and Chainpur. The Mahakali and Seti rivers are being dammed and diverted for the purposes of hydroelectric power generation and irrigation. These developments offer possibilities for agricultural expansion. Finally, the frontier postings have become important trade and transport areas. Such changes ensure a wider accessibility for the region and greater prospects for regional development.

Elsewhere in Nepal, significant gains in the amount of area under cultivation are found mainly in the southern foothills of the tarai region. In some instances, in the tarai the expansion of farmland at the expense of the forested areas is truly outstanding. In the Nawalparasi district in the central tarai, for example, the annual rate of increase of crop area exceeds 10 percent a year. After the eradication of malaria in the tarai in the mid-1950s, the southern lowland region opened to immigration. Both planned and illegal settlements have in turn led to the expansion of farms in the most fertile regions. Between the 1960s and 1970s, almost one-fourth of the tarai forests was converted to farmland.[11] During the past decade, migration to the tarai from elsewhere in Nepal and from India has increased considerably, leading to continued high rates of farmland expansion.

In the eastern Himalaya, meanwhile, Darjeeling reports a high rate of loss of crop area, while Bhutan exhibits a significant gain. Over the past several decades, the Darjeeling region has become increasingly urbanized and the most productive farming lands there are now devoted to cash crops, including the important tea plantations. Bhutan, on the other hand, seeks to increase its measure of food self-sufficiency partly by the extension of agriculture into formerly uncultivated regions. The government-sponsored land settlements and the spontaneous conversion of forest land to agricultural use during the past thirty years have contributed to the increase in crop area in that country.[12] Bhutan, however, recognizes that only small additions in agricultural production are possible by developing unused land, and the country envisions instead a more integrated approach to agriculture, one which combines horticulture and cereal agriculture with broader land and water conservation needs.

Insufficient information exists to track the changes in crop areas during the contemporary period in Sikkim and in the far eastern regions of the upper Assam Valley. In Sikkim, where approximately one-tenth of the land is devoted to agriculture, food security is less of a problem. Household landholdings and per capita consumption of food grains there are higher than in many other parts of the Himalaya, and future agricultural development will likely center on horticulture and commercial farming as well as on food-grain imports. Among the far eastern Himalayan regions, shifting cultivation remains a significant form of agriculture. The mobility of such farmers and the volatile nature of land use makes it difficult to assess changes in the agricultural landscape. Nonetheless, in places that are most accessible to roadways and to the river valley travel routes, the expansion of cultivated land is on the rise.

Changes in the Himalayan farm area result from population shifts, from social aspirations and material demands, and from development infrastructures such as roads, irrigation, and markets. Population shifts, in particular, show the direct pressure of people on the agricultural resource base. A simple measure of this pressure is the physiographic density of population in crop areas. Such measurements have been made for the Himalayan districts in 1960 and 1980 to allow a comparative study of change.

With the exceptions of the southern extensions of the Tibetan plateau, where new farm areas are being developed as a result of the extension of irrigation systems, and the piedmont plain, where lowland forests continue to be cleared for farmland, little opportunity exists to expand farmholdings.[13] Meanwhile, the number of persons who depend upon existing areas of cropland continues to increase, leading to high population densities in many farmland areas and to the host of land degradation problems identified in the earlier chapters of the book. The growing demand for food has put a great deal of pressure on those ecological processes that maintain the mountain habitats. Additionally, the inability of the farming systems to match the human food needs has led to human impoverishment, to dietary restrictions, and to outmigration as people seek livelihoods elsewhere. These are some of the more obvious human dimensions of environmental decline.

The crop-area population densities computed and mapped for 1960 show a mainly stable situation across much of the Himalaya with the exceptions of the northwestern region and Nepal. A few high-density areas were located in the northwest. In Ladakh, where less than 2 percent of the total area was farmed in the 1960s, the irrigated land under cultivation is the main reason for recent production of cereal grains. Otherwise, the farming systems in Ladakh and in the neighboring arid regions of Zanzkar and Spiti rely heavily on yak and sheep grazing to produce food and income for the scattered villages.

Low to medium farmland population densities were discovered in the Garhwal and Kumaun regions in India, indicating only slight pressure in these two places from human population on crop area prior to 1960. In Nepal, however, thirteen districts showed very high settlement densities. All but two of the districts were located in the middle-mountains zone. An additional nineteen districts in Nepal reported medium to high densities in 1960. Those areas accounted for almost two-thirds of all of Nepal's districts for which 1960 data was available.

The widespread high and very high crop-area population densities reported in 1960 suggests that the lack of adequate farmland has been a problem in Nepal for quite some time. The high rates of outmigration from the mountains to elsewhere in the country and abroad,

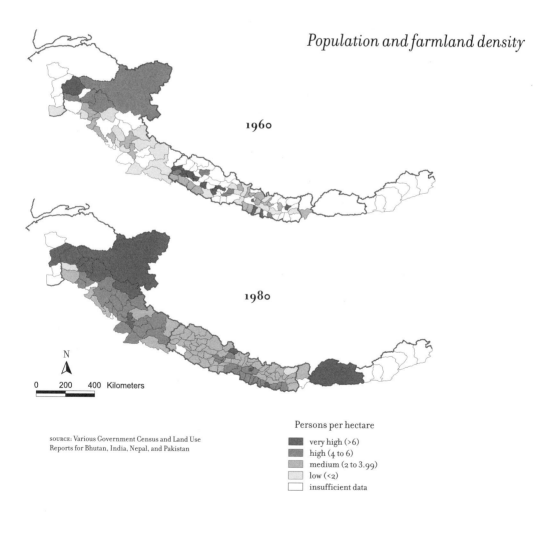

Population and farmland density

1960

1980

N

0   200   400 Kilometers

SOURCE: Various Government Census and Land Use
Reports for Bhutan, India, Nepal, and Pakistan

Persons per hectare

■ very high (>6)
■ high (4 to 6)
▨ medium (2 to 3.99)
▨ low (<2)
□ insufficient data

The population of the Ziro Valley, which is a tributary of the Brahmaputra River in the far eastern Himalaya, has grown considerably in recent decades. Photo by P. P. Karan

which began in earnest in the 1960s, further attest to the pressures of rising human demand on limited agricultural land. When the scarcity of land is coupled with low productivity from land degradation or from the inclusion of infertile steep slopeland into the farm systems, the agricultural landscapes in Nepal become some of the most severely affected areas in the entire range. It is reasonable, therefore, that the common images of a looming Himalayan crisis should be so alarmingly drawn; they derive, after all, from firsthand observations of the Nepalese countryside.

The farmland population densities computed and mapped for 1980 show a steady worsening of conditions across much of the range. This period corresponds to the most intensive phase of resource development in the Himalaya.[14] The largest increases in densities occurred in the western mountains of Garhwal and Kumaun, where by

1980 each hectare of cropped land supported on average almost five people. This situation compares most unfavorably to that of 1960, in which each person had close to a half hectare of farmland. In Nepal, which already showed a great deal of human pressure in 1960, the land situation has steadily worsened such that by 1980 the entire country shows medium-high to very high densities. Even in the tarai region, which only opened to agricultural settlement in the 1960s, most districts by the 1980s had high farmland population densities.

In the far eastern Himalaya, approximately 10 percent of Sikkim is cultivated (almost one-third if the high-elevation regions are omitted), and each household has access to about one and a half hectare of farmland. This compares favorably with the highly stressed regions of the Himalaya. Nonetheless, there is a slight shortage of food grains in Sikkim which is met by food imports. The lack of reliable land and population measurements in Bhutan precludes a careful computation of farmland pressures there.[15] Using the best available government figures, the average crop-area population density for the entire country ranges between two and five persons per hectare of tilled land. With a population growth rate at more than 2 percent per year, Bhutan is faced with growing demands on food production. The relative scarcity of arable land makes agricultural innovations tantamount to that country's development goals.

Among the far eastern regions of the upper Assam Valley, only loose estimates of cultivated land and population are available. They show, however, that this region has been the least affected by farmland pressures. Much of the agriculture practiced by the hill tribes is shifting cultivation, and the overall low population densities recorded in Arunachal Pradesh point to low pressure on existing cultivated landholdings. Where migration has fueled population growth, such as along the fertile lower stretches of the Subansiri and Siang rivers and, notably, in the Ziro Valley, localized stress is reported in the farmholdings. The region as a whole still shows fairly stable agricultural conditions, but the trends in those areas suggest additional problems in the near future.

The broad picture of unstable farming systems in the Himalaya, at least in respect to population demands, is matched by the alarming decreases in harvest yield and soil productivity reported across the range. The amount of farmland per person declined by more than 30 percent between 1960 and 1990. The average yields also declined, although at a slower rate. In Nepal, for example, cereal-grain yields dropped more than 15 percent in the 1980s. The greatest decline occurred for maize, which is a vitally important dry-land crop grown throughout the Himalaya. Under such conditions, the gap between food needs and availability will no doubt increase.

These trends also indicate the general failure of many mountain farming systems to keep pace with the rising tide of human food needs. They raise the critical issue of food security for future generations. In addition to the conservation efforts aimed at stabilizing the loss of forests, there exist a widespread need for structural adjustments in Himalayan agriculture and a renewed attention to the development of new opportunities for more intensive use of existing farmland.

Until now, most agricultural development has been driven mainly by the demands of the market economy for greater production of high-value specialty crops such as fruits, tea, vegetables, and spices. It is clear, however, that future efforts need to be aimed at increasing food production for the villages. A number of human interventions are possible to facilitate such changes. A reinvestment is needed, for example, in the traditional conservation strategies of mountain societies, so that the recycling and nutrient transfer potential of indigenous, organic-based agriculture is enhanced. Market forces can be harnessed which prioritize food security for local inhabitants over corporate profits for distant agribusiness. Public-sector investments in new farming technologies are possible, especially those that extend rather than displace traditional farming practices. In sum, the future course of agriculture in the Himalaya must respond simultaneously to the degradation that has already occurred in the land systems as a result of overintensification of resource use and to the food-security problems that loom on the horizon as one of the region's major obstacles to sustaining livelihood.

Pressure of Population on Forests

The actual rate of forest loss in the Himalayan region remains uncertain despite the many studies that deal with land-cover changes in local settings.[16] The absence of regional assessments is due in part to the difficulties in obtaining the historical as well as the current areal

Food aid for Nepal arrives by train at the Raxaul roadhead, located along the India-Nepal border. No railways penetrate the Nepalese interior, so food shipments intended for the mountain regions still require human porters. Photo by P. P. Karan

measurements of forests. It is estimated that about 1 percent of the Himalayan forests are depleted annually through clear cutting. That loss is compounded, though, by the degradation that occurs daily in the forests as small trees, branches, leaves, and other items are removed for subsistence needs as well as for commercial purposes. Overgrazing by village livestock, excessive fodder and fuel-wood collecting, and the annual fires set to burn pastures all take a heavy toll on the forests. Plant gathering, especially for medicinal products, and charcoal making are additional sources of human pressure on the forests.

Some broad-scale regional variations in forest conditions exist across the Himalayan range. In the western region, much forest depletion has occurred during the past three decades. That is due in part to the rapid population growth—fueled among the Indus districts of Pakistan, for example, by migrations and refugee movements—and to the increased commercial exploitation of forests. In the Mansehra district in Pakistan, the illegal felling of trees by timber merchants is widespread along the new feeder roads of the Karakoram Highway. Similar conditions prevail in Kashmir, where government forest regulations are not enforced because of the current political climate and civil violence. The per capita forested area in the mountains of the Indus area declined

A heavily forested area located in the middle mountains of Bhutan. Photo by P. P. Karan

from one-sixth hectare in the early 1970s to one-tenth hectare in 1990. These rates are among the lowest in the entire Himalaya.

The forest situation is somewhat better elsewhere in the western mountains. The per capita forest area is two-thirds hectare in Garhwal and one-half hectare in Kumaun. The Kulu Valley in the state of Himachal Pradesh shows good forest conditions remaining, with a per capita forest area of almost one hectare, despite recent intensive commercial developments along the Beas River, especially near the town of Manali. In those places, deforestation is increasing quite rapidly as tourism, orchards, and other commercial land uses expand in the valley. Much of the contemporary forest loss in the Garhwal and Kumaun regions occurred between 1960 and 1980 when population and commercial pressures were at their highest. The Indian Forest Conservation Act of 1980, which extended government control over forest use, has contributed to a notable decline in deforestation there. Nonetheless, where the controls are not enforced, illegal encroachments of the forests are still quite common.

The natural habitats in Nepal, on the whole, remain some of the most threatened ecosystems in the entire Himalayan region. The Himalayan environmental degradation theory was based upon observations of forest conditions in that country. Official estimates in the 1980s showed the overall rate of forest loss to be somewhere between 2 and 5 percent per year. More recent estimates place it at around 1 percent per year, matching the regional average and suggesting an improvement over the earlier trends. But the new reforestation initiatives have not kept pace with the earlier rates of forest loss—less than 5 percent of the area deforested has been planted in trees.[17] The outright depletion of forests in Nepal, coupled with the growing human pressures in the landscape, result in watershed deterioration and soil erosion in many parts of the country.

Despite national trends, marked regional differences exist in the status of the forests in Nepal. The kingdom shows an average per capita forest area to be a very low one-fifth hectare. In the western mountains, however, the per capita forest area is eight-tenths hectare, which compares favorably with other parts of the Himalaya. The central middle mountains of the kingdom, with a per capita figure of one-tenth hectare, is the most seriously degraded forest region in the country. It is the region that is most often cited in the ecocrisis literature, being easily accessible from Kathmandu, but is not necessarily indicative of conditions elsewhere.

Much of the effort to reforest Nepal takes place in the central part of the kingdom, especially in the Sindhu Palchok and Kabhre Palanchok regions located near Kathmandu Valley and in the hills directly north of the town of Pokhara. Long-term investments in forest regeneration are coordinated there by various international agencies. The Nepal-Australia Forestry Project, for example, reports quite favorable forest conditions in the Sindhu Palchok project area. That project is widely cited as an example of successful forest intervention in Nepal. But the large financial and institutional investment in such a small area is not likely to be replicated elsewhere in the country. It remains to be seen, therefore, if the successes of these programs can be transferred to other parts of the country.

The forests of Bhutan, meanwhile, are some of the finest natural areas left in the Himalayan range. The population densities in Bhutan are generally still quite low, and the levels of commercial exploitation of forests remain fairly modest. Bhutan's large forests are thought to be vital to the biodiversity of the entire Himalaya. Nonetheless, deforestation occurs in Bhutan at an estimated rate of 2,000 hectares per year. The main reasons for the forest cutting are the expanded commercial felling that occurs alongside the roadways, the open grazing by cattle in the forests, and the village demands for farmland, fuel wood, and fodder. Where concentrations of human and livestock populations occur, such as in the eastern and southern regions of the kingdom, the degradation of the forests is most severe.[18] The average per capita forest area in Bhutan is two and a half hectares. That is still the highest value in the entire Himalaya, with the possible exception of regional forests in the far eastern Assam highlands. Forest survey information is unavailable in much of the eastern sector. The inhabited forests in Arunachal Pradesh still are mainly used for shifting cultivation, locally called *jhum*. The field rotation cycle has been radically shortened in recent decades, though, from fifteen years to five years, which does not permit a proper regeneration period of the forests[19] and is causing problems of deforestation. The recent stripping of forests in Arunachal Pradesh has also been accomplished by sawmill barons and illegal loggers who, despite strict regulations on cutting and exporting timber, still manage to cut the forests for private gain. Logging permits given to the tribal people for subsistence purposes are traded on the black market to commercial operators. The logs are floated down the tributaries of the Brahmaputra River and recovered in Assam, where they are then hauled by trucks to the market.

Without better management of the forests, more widespread land degradation is likely in the mountains north of the Assam Valley. The overall contributions of population growth and subsistence demands to deforestation must be weighed against the history of commercial forest exploitation in the Himalaya if a better understanding is to be gained about landscape change in the mountains. Still, the measure of forest change against population change is a useful indicator of regional forest dynamics. The association shows in part why forest change occurs, but even more importantly it indicates how accessible the forests are to the villagers. That is a critical feature of rural life throughout the mountains. The forests provide much of the sustenance of life—food, fuel, medicine, fodder, building materials—and so they need to be close at hand.

Forest Change

Drawing upon government documents and census reports, we assessed forest-area change for 120 Himalayan districts during the historical (1890–1950) and contemporary (1960–90) periods. They show a wide range of circumstances. During the historical period, high or significant rates of forest loss are recorded for eight districts in the Indian western Himalaya, including parts of Garwhal, Kumaun, and the mountains in the Indus region. The rates of forest loss during the historical period averaged a little more than 2 percent per year. The districts reporting the highest rates are those where significant forest extractions took place during the colonial period or where urban growth began at an early period. At the same time, though, and among the adjoining districts, some gains in forest area were also recorded. The Chamba, Kulu, and Mandi districts located along the Beas River, for example, showed significant advances in forest areas during the historical period. It may well be that such additions are merely statistical anomalies or census discrepancies, since little anecdotal evidence exists to support the notion of an actual expanding forest base during the historical period in the western mountains.

Colonial records kept for the western Himalayan region provide a fair assessment of forest conditions there, but historical information for much of the rest of the mountains is scarce, prohibiting a full comparison of regional conditions throughout the Himalaya. The historical land-use trends described in earlier chapters suggest overall losses of forest cover across much of the central range. The mountains in Arunachal Pradesh, meanwhile, remained mainly forest covered until only recently. During the contemporary period, better information is available for most of the range. Between 1960 and 1990, about one-third of the Himalayan districts reported a forest loss. While that is significant, it is not as bad as the reports of the 1970s and

1980s predicted: an almost total denudation of the Himalaya within a matter of decades. Moreover, much of the forest loss occurred in the southern tarai and outer foothills zones. That corroborates local studies that show the increasing pressure placed on the lowland forests by timber merchants and immigrant farmers during the 1970s and 1980s.

But fully one-quarter of the Himalayan districts reported a gain in forest area during the past few decades, with much of the remainder showing relatively stable forest conditions. Such findings require a modification of the regional deforestation scenario. Even in central Nepal, where forest conditions are among the worst in the Himalaya, the forests are so diverse and the landscape dynamics are so fluid that a countrywide estimate does not portray well the actual suite of environmental problems occurring there. Generally, forest conditions in the western part of Nepal have remained good, although eight districts now report serious deforestation. The highest rates of forest loss in western Nepal occur among the districts that border India and those in the remote Dolpo district, where annual forest loss is measured at 3 percent a year. The highest rates of forest gain in Nepal, meanwhile, occur in the eastern middle mountains. In a broad west to east swath, from the Kathmandu Valley to the Sikkim border, we find a tier of seven districts that exhibit an average forest gain of more than 4 percent per year. That is an encouraging phenomenon and one that flies in the face of most media accounts. The highest rates of forest loss in Nepal, as expected, occur in the tarai, where immigration and land clearing are widespread. For example, the Bardiya district in the western tarai reports an annual deforestion rate of greater than 2.6 percent, Chitwan in the central tarai reports 2.2 percent, and Saptari in the eastern tarai shows a 2.8-percent forest loss per year.

Frequently, districts reporting gains and those reporting losses actually adjoin one another. That paints a disconcerting and seemingly contradictory picture. It may be a problem of unreliable statistics, but the reported variability does reflect a large number of factors: the amount of land under forest cover, the size of the human population, the type of economy, and the level of infrastructure development. The overall complexity of forest dynamics at the district level throughout the Himalaya caution against any regional generalizations, as least for the time being, in the mountains.[20] Nevertheless, serious levels of forest loss do exist in the range, compelling a number of much-needed conservation programs in specific places.

Forest-Area Population Density

The relationship between population change and forest area is critical for understanding land dynamics in the Himalaya. The per capita forest figures presented above show representive forest-area population density values. They indicate the level of demand that the human population exerts on available forestland and the accessibility

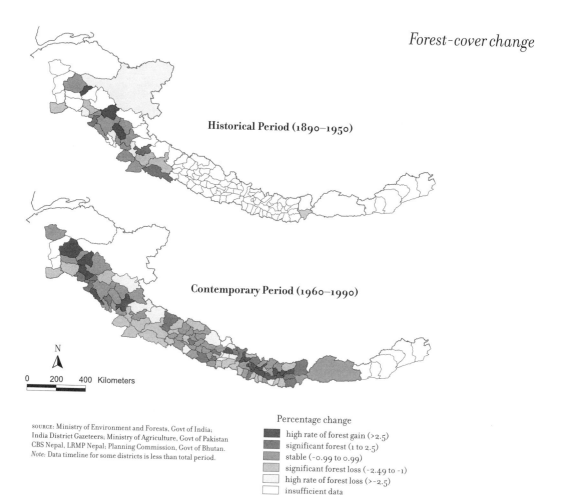

Forest-cover change

Historical Period (1890–1950)

Contemporary Period (1960–1990)

N

0   200   400 Kilometers

SOURCE: Ministry of Environment and Forests, Govt of India; India District Gazeteers; Ministry of Agriculture, Govt of Pakistan CBS Nepal, LRMP Nepal; Planning Commission, Govt of Bhutan. *Note:* Data timeline for some districts is less than total period.

Percentage change

■ high rate of forest gain (>2.5)
■ significant forest (1 to 2.5)
■ stable (-0.99 to 0.99)
■ significant forest loss (-2.49 to -1)
□ high rate of forest loss (>-2.5)
□ insufficient data

of forests for cultural purposes. We computed the density values for all the Himalayan districts for the years 1970 and 1990 in order to capture some of the change that may have occurred between human numbers and forests during the contemporary period.

The forest-area population density values are important not because they necessarily imply deforestation. In fact, the wholesale clearing of forests has slowed considerably in many parts of the Himalaya despite increases in population densities. Rather, the ratios indicate the possible pressures of the subsistence economy on the forests. Whether degradation occurs is largely the result of the presence or absence of strong local conservation measures. The major characteristic of landscape change nowadays is the conversion of canopy forest habitat to more open shrublands and grasslands.

The shift occurs not as a result of tree cutting but because of more intensive uses of the forests for foraging and livestock grazing. Those practices do not necessarily remove trees, but they do destroy the understories. The removal of biomass from the forests beyond the level at which the forests can regenerate will eventually reduce the growing stock of the forests and contribute to cycles of degradation even as the existent trees remain in the landscape. Given villagers' need for fodder and fuel wood, the low per capita availability of forests suggests critical areas of environmental stress. Despite many studies, the actual rates of fuel-wood usage and the average fodder needs of village livestock remain elusive figures.[21] Consequently, it is still not possible to determine with any great confidence the true regional impact of human numbers on forests. Again, that impact is largely dependent upon the social and environmental conditions that prevail in particular places.

The trends in population growth and forest change show some distinctive geographical patterns of forest-area population density. In 1970, the pressures of population were greatest on the forests in those areas of the middle mountains and outer foothills where commercial developments, infrastructure expansion, and urbanization are most advanced. Those things produce high demands for forest products and add to the local population by promoting additional migration into the region. In the southeastern part of Nepal and in Darjeeling we also find high-density values for much the same reasons. No accurate data is available for 1970 for Bhutan or for the far eastern Assam region which would allow comparable assessments. Field visits to Bhutan and Arunachal Pradesh show, however, that the recent trends in commercial development and road construction there have created corridors of intensive land use and forest extraction.

Between 1970 and 1990, a gradual intensification of population pressure on forest land occurred in the Himalaya. The pattern is most pronounced along the southern tier of the outer foothills and piedmont plain, where some districts in India report per capita forest areas of less than 0.03 hectares. According to the most recent figures in Nepal, the average per capita forest area in the tarai during the early 1990s was 0.07 hectares. That is an overall decline from 0.13 hectares

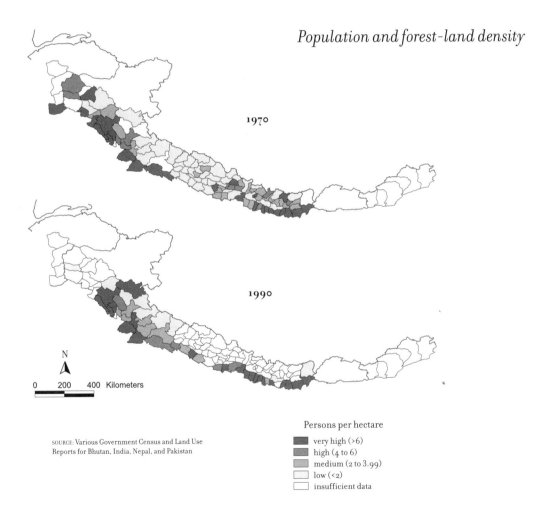

Population and forest-land density

1970

1990

N

0   200   400   Kilometers

SOURCE: Various Government Census and Land Use
Reports for Bhutan, India, Nepal, and Pakistan

Persons per hectare

■ very high (>6)
■ high (4 to 6)
▨ medium (2 to 3.99)
□ low (<2)
□ insufficient data

The cultural landscape in the valley of the Sanghla River shows a mix of forest, pastures, and farmland near Chitkul village. Photo by D. Zurick

in 1970. While reliable forest data may be scarce for many of the mountainous districts of Nepal in 1990, some improvements have taken place where out-migration and reforestation schemes contribute to more balanced population-forest ratios. The most favorable per capita forest values occur, as expected, in those districts where population has stabilized or declined and where forest area has increased. Forest-area data for 1990 remains sketchy for Bhutan and for the far eastern region. It is likely, however, that with continued population growth and declining forests, density values will continue to rise in those places. Overall, the most striking changes in land cover take place along corridors of industrial activity, which, in turn, follow the new roadways. That has contributed additional edginess to the Himalayan landscape.

The Demands of Survival

To relate population numbers to forest areas as a means to predict general environmental stress assumes that some fairly consistent rate of forest-products use exists in the villages. That is not the case. The use of forest fuel wood, food, fodder, and medicine varies widely instead across the environmental zones and culture areas in the mountains, and accurate measurement is difficult, perhaps impossible. It is noteworthy, however, that the most important sources of pressure on the land derive from a combination of subsistence and commercial economies. The balance between the two may tell a lot about the sustainability of rural societies, particularly as they move toward more industrial forms of life.

The primary requirement of the mountain villages is farmland, and the conversion of wilderness to domesticated land is a necessary component of life in the mountains. Under conditions of fairly low population pressure only the most fertile and productive mountain lands are cultivated. Those may include the bottom lands along rivers which provide terraces for rice farming, as well as the upper slope lands that support dry-land grain, fruit, and vegetable crops. Owing to the organic structure of Himalayan agriculture, most farmers are keen to maintain forests nearby for nutrient cycling purposes. The most extensive forests generally occupy the least productive land, often growing on steep slopes and ridge lines, where they play important roles in providing natural habitat, in producing biomass for village use, and in stabilizing the slopes.

It is only with severe population pressure that such forests are likely to be converted to farmland. Even then, under the carefully managed terrace systems found across the Himalaya, the conversions need not necessarily lead to land degradation. In fact, throughout the mountains we find an incredibly rich and sustainable landscape created by humans. Sophisticated local engineering networks have carved hill slopes into terraces that distribute irrigation water through a vast network of interconnected fields. These same terraces work to

Heavily used grazing lands in an upper village pasture in western Nepal. Photo by P. P. Karan

prevent soil erosion. Complex social institutions guide the allocation and use of land and water resources according to household and village ideals.[22]

However, when the traditional land systems break down, or where the pressures for food compel more intensive land clearing on steep slopes, then the potential arises for serious environmental damage. Changes in agricultural practice that commonly lead to degraded conditions on the mountain farms include the reduced time given to fallow land and crop rotations, the cultivation of marginal land, the poor maintenance of hill terraces, and the shift in crop patterns from diversity to monoculture.

Much of the attention on environmental degradation is placed on how villagers use their forests. The clearing of wildlands for farmland expansion is an obvious use of forests. But villagers rely on the forest habitats for a wide range of needs. The forests and farms are linked by the transfer of nutrients which occurs when fodder is collected and livestock are allowed to graze among the trees. The dung that is produced is a key additive to soil fertility in the farms. Hence, the maintenance of forests as repositories of fodder for livestock greatly determines the sustainability of organic practices on the mountain farms.

The forests also provide fuel wood for heating and cooking and for numerous local processing industries such as drying or distilling food, pottery making, and tool foundries. While various estimates of fuel-wood use appear in the Himalayan literature, on the whole nearly 80 percent of the energy used by villagers still comes from firewood obtained in the forests.[23] Where the forests are degraded and fuel wood is scarce, shortage of energy is a critical concern. In order to lessen the impact of rising energy demands on mountain forests, numerous efforts are under way to provide energy alternatives. They include the renewable forms of hydropower, photovoltaic panels, and biogas plants, as well as the more conventional sources of kerosine, petrol, diesel, natural gas, and coal. Most of the alternative energy sources are expensive and require the villagers to move beyond subsistence resource environments into the fledgling market economy.

The villagers also rely on the forests for a host of food, medicinal plants, and other nontimber products. Wild grains, fruits, tubers, mushrooms, and vegetables are commonly collected from the forests,[24] and provide important additions to the simple diets of rural people. The range is well known for its medicinal plants, many of which now have a high commercial value and are gathered for export trade, much of it illegal. Trafficking in medicinal herbs is epidemic now in the mountains and threatens the sustainable supply of such items for use by the villagers.

To a large extent, the conditions of environmental degradation in the mountains reflect the poverty in peoples' lives there. The hazards of life are mirrored in the quality of the immediate landscape. The desperation that drives people to use resources badly also puts people at greater risk. It threatens the integrity of the mountains both as a place of human inhabitation and as one of earth's great natural places. Land degradation in its most human context heightens environmental risk by lowering the security of production from fields and forests.

Poverty and Environmental Risk

THE ROOTS OF POVERTY in the Himalaya lie in the historic appropriation of the region's natural wealth by outside interests; in the limited access of villagers to farmland and other key resources; in population growth and land degradation; in the concentration of power among the rural elites or distant urban centers; in the fact that Himalayan people occupy the geographical margins of state and world political economies; and in the constraints on productivity that are imposed by the unique conditions of the highland environments. Such factors together contribute to the risks and burdens of human life in the mountains.

Dire Straits: The Conditions of Poverty

In most global assessments of economic performance, the Himalayan countries rank at or near the bottom of the world poverty lists.

The United Nations Development Program annually classifies countries according to income level and potential for human development; the Himalayan region consistently scores in the lowest division, above only the most impoverished countries of Sahelian Africa. The per capita incomes of the Himalayan countries are among the lowest in the world—$105 U.S. in Bhutan, for example, and $190 U.S. in Nepal. Overall, the Himalaya per capita income averages $157, compared against an average income of $970 for the developing countries. In the western mountain districts of India, over one-fourth of the population lives below the Indian poverty line of 6,380 rupees (less than $200 U.S. in 1998 equivalents). In most cases, the income levels in the Himalaya have dropped in recent decades, so its economic picture is even more bleak.[1]

The use of such statistics is limited by the usual problems of unreliable and conflicting data.[2] There simply are few accurate tallies of economic conditions in the mountains, so we are left to piece together often wildly divergent accounts. The national figures also tend not to include accurate reports of foreign earnings or the black-market economy that results from illegal activity. Aggregate measures also fail to show how the reported national earnings are distributed among the diverse populations. There are extremely wealthy people living in even the poorest countries. Finally, they do not capture well at all the output of the traditional economies. The latter point is particularly important in the Himalaya, where about 70 percent of the total population is employed in subsistence agriculture. In most places, almost everyone is still fully engaged in farming. In Bhutan, about 95 percent of the people are farmers.

The off-farm employment that occasionally supplements agriculture generally goes unreported. Involvement of rural people in the formal market economy, which constitutes the reported aggregate economic statistics, remains relatively minor throughout the Himalaya. These factors add to the uncertainty of computing wealth. A better measure of poverty under such circumstances would explain not so much the workings of capital, but the productivity of the natural environment. When environmental productivity falls or when access to sufficient productive land is curtailed, the real poverty of the mountain communities increases sharply.

According to the major indices of economic development, the

earnings of the Himalayan region are limited by numerous economic and social factors. The sluggish output of the rural economies combined with the high rates of population growth, the stagnation or slowness of industrial sectors, the overall lack of transportation or other important infrastructures, and the low levels of national export trade combine to suppress economic growth. When we consider, however, the conditions of poverty in the Himalayan subsistence context, a different set of factors comes into play. They include the various forms of cultural disruption, social inequity, and economic alienation; the rapid population growth and the breakdown of subsistence economies; and the accelerating commercial development of the Himalayan worlds. Taken together, these alternative indicators of poverty add greatly to the environment risks of the already impoverished mountain communities.

The imperial efforts of the early Rajput hill kingdoms in the western and central Himalaya and the later British interventions in the Garhwal and Kumaun regions disrupted many of the traditional systems of culture and their deeply embedded relations to the land. That is not to say that the traditional cultures in the western mountains were solely dependent upon agriculture for their livelihoods. Rather, throughout the Himalaya, the rural communities have relied on a diverse base of farming, petty trading, and occasional wage labor to meet their basic needs. These practices more or less augmented the farming and livestock activities. They were grounded in long-standing cultural relationships that often were ignored by the foreign rule over the mountains.

The modern tendencies of national development, likewise, frequently ignore the traditions of land tenure and resource trade and serve to socially distance the household from farmlands and forests. Government restrictions on indigenous use of the forests, for example, have curtailed the already limited options for agriculture as well as the potential for local profit gained by trading in various forest products. When coupled with the lack of off-farm opportunities, the new forest rules aimed at commercial timber management add greatly to the measure of poverty in the Himalaya and severely reduce the security of the peasants' access to natural resources. Often the only recourse that is open to the villagers is coordinated environmental activism.[3]

Freedom from historical imperialism has not necessarily improved the environmental security of the Himalayan people.[4] Many of the national efforts to modernize basically continue the old policies of centralized control over land resources and further restrict the rights of indigenous peoples by imposing high licensing fees for resource use under the threat of heavy fines. The terms that are now used to describe government policies in the mountains may be new—investment priorities, resource allocation, factor/product pricing, and so forth—but the primary intention is to further exploit their commercial potential. Against such efforts, the level of poverty in the Himalaya unfortunately continues to increase.

The unique conditions of the mountain environment—fragility, marginality, and diversity—are successfully integrated in many of the Himalaya's cultural adaptations. The traditional economies exploit manifold environmental niches, employ organic methods of farming, and foster social linkages that favor equitable and long-lasting economic exchanges. These often result in quite sustainable lifestyles in which poverty is kept at a minimum. Under the new conditions of modernity, however, in which demands on resources have increased and the cultural traditions governing the use of resources have deteriorated, the levels of poverty in the mountains increase. Moreover, the commercial policies that are proposed to alleviate poverty may add to regional and social disparities in wealth and levels of environmental security. The current industrial model of economic growth, which has advanced national development elsewhere but with considerable environmental cost, will inevitably deepen the uncertainty of mountain livelihoods.

The Burden on Women

It is said in the Himalaya that women hold up half the sky. They receive, however, only a minuscule portion of the wealth produced by the land. Alleviating poverty for women and children in the mountains is one of the region's greatest challenges. A major factor in determining the distribution of wealth in the Himalaya is land ownership. Because of the nature of local inheritance practices, or the lack of political clout, women are most prone to landlessness and hence are most vulnerable to poverty.

Women, on the other hand, are primarily responsible for securing fuel wood, water, food, and other basic provisions.[5] The time women spend on such daily chores as firewood and fodder gathering, foraging, and water collecting greatly surpasses that of men. In the subsistence economy, these contributions are essential but receive no remuneration. Without money, women cannot avail themselves of the development opportunities open to men who work for cash as wage laborers in the market economy. Hence, women are disadvantaged in both the subsistence and commercial sectors of the developing Himalayan economies.

The linkage between environmental degradation and increasing levels of poverty in the mountains also most severely affects village women. Resource scarcities require them to spend more time gathering forest products, tending fields and livestock, and collecting water. In effect, the households cope with ecological stress by utilizing more female labor.[6] Thus the greatest burden of land degradation is placed on the shoulders of women. It is not surprising, therefore, that they should lead many of the environmental resistance struggles in the Himalaya. The Chipko Protest and the opposition to Tehri Dam are two outstanding examples of community activism led by women. All this makes clear the fact that any efforts to combine environmental conservation and poverty alleviation in the mountains must include the crucial support and commitment of women.

The Alleviation of Poverty

The primary scope of programs aimed at removing poverty in the Himalaya is threefold: extend state welfare to mountain people; integrate mountain economies with regional and national economies; and develop the natural wealth of the mountains. The state welfare program includes numerous efforts to provide health and education opportunities to the rural population; nevertheless, the levels of access to health care and the rates of literacy in the Himalaya are among the lowest in the world, accounting in large part for the overall low human development indices reported by the United Nations Development Program.

The state welfare policies also include direct government subsidies to agriculture. They provide credit and technology transfers, mainly to a minority of wealthy farmers. Perhaps the most notable type of direct welfare payment in the mountains is the distribution of relief food supplies in needy regions. In the northwestern Indian Himalaya and the western districts of Nepal, such provision has been growing steadily during the past several decades. Despite these efforts, the prevailing high rates of national poverty limit overall the services that the Himalayan countries can provide to their people.

Regional integration programs meant to relieve poverty follow along the predictable lines of improving transportation and other infrastructures, developing economic markets, and promoting decentralized government administration. Their activities in fact cornerstone most current regional development schemes in the Himalaya. They tend to link poverty in the peripheral areas with the lack of economic opportunities. Hence they intend to promote market activity, which presumably will, in turn, provide jobs. Characteristically, such efforts serve best to shore up the influence of the political centers in the rural areas and to further marginalize the traditional practices of local people. Their efficacy in reducing poverty thus may be incidental at best.

The development of regional markets, for example, tends to focus on a single element in the mountain farming systems, which are in fact composed of carefully integrated practices. For example, horticulture, livestock, forests, and farms are considered separately in accordance with related economic sectors, but in reality they are organically linked under traditional land uses. The agricultural scientist N. S. Jodha has called this flawed development approach, which is borrowed from the plains economy, the "missing mountain perspective."[7] Such interventions are criticized because they tend to disregard the unique circumstances of the Himalayan environment and to promote single-crop farming rather than the diverse blend of crops that describe traditional mountain farms. It is feared that such programs may actually heighten the level of risk among marginal mountain communities.

A major effort toward regional integration is under way in the Himalaya in regard to tourism development. Mountain tourism is proposed to be a key economic activity under most sustainable-development programs. The movement of tourists across the mountain landscape, the provision of tourist services, and the employment of people in the tourism industry are thought to indicate real possibilities for tourism development in the Himalaya.[8] The amount of tourism earning and its distribution among rural areas are generally far less, however, than what the planners hope, while the adverse environmental and social consequences of tourism may be quite pronounced in the landscape.

Under the conditions of declining farm productivity and degraded natural habitats, the provision of economic alternatives is crucial in order to resolve some of the region's poverty problems. In some cases, this may be possible through the development of off-farm employment that is compatible with local subsistence needs. A variety of efforts aimed at developing agroprocessing industries, nontimber forest production, and handicrafts are proposed in addition to the possibilities of tourism. But the most common strategy for generating wealth in the mountains remains the extraction of basic natural resources. In most cases, the wealth thus produced does not stay in the mountains. It is intended instead for corporate holdings or for political gain. The consequences of such activities, however, usually remain in the villages and make life there even more difficult.

Developments of water resources for hydropower generation, of farms for horticulture and market crops, of forests for timber, charcoal, and medicinal herbs, and of the mountain landscapes themselves for parklands and tourism expansion are primary examples of the big commercial programs. We look at some of those developments in the next part of the book. They predict continued marketing of the Himalayan worlds. Modernity, in such cases, most often strengthens the inequities that already exist between the rural people in the mountains and those living in the industrial societies of the plains and towns. In this case, it may lead to the further alienation of highland people from their land. And it contributes to a path of change in the mountains that foretells continued problems of environmental sustainability and social equality. Against its programs are the continuing struggles of local people to retain their rights over the mountain landscapes.

PART FOUR

THUNDER IN THE mountains

A vegetable market in Nepal. Photo by P. P. Karan

The Pale of Modernity

THE HALLMARK OF CONTEMPORARY Himalayan landscapes is their rapid transformation, but then change describes all places, including the most distant mountain valleys. In the Himalaya, the cycle of geology continues in such a fashion that the land itself is changed. The social history of the mountains, meanwhile, has been wrapped up in the successive advances of empires and the struggles of native people. Now the modern world is ready to climb the summits. We may think at first of simple markers in the landscape: satellite dishes bringing Western broadcasts to the remote stone and mud villages of Ladakh, tinned food labeled with foreign scripts appearing on the shelves of distant caravansaries in Zanzkar, cellular phones ringing atop Mount Everest, power lines spanning the glacial valleys in Baspa and Spiti. For some people, those additions are disconcerting features in the landscape, but in fact they intimate much more than

12

puzzling appearances; these innovations foretell new identities for remote places.

The change that is blowing in the mountain winds may be most evident to those who find their livelihoods threatened by farms that have been rendered unproductive by soil erosion or by government policies that restrict access to forests and fields. In such cases, peoples' homes can no longer provide them with even the most basic needs. Change is also inherent in the experiences of young people living in the Himalaya, whose exposure to Western ways compel aspirations that may not be met by their cultural past. These youth seek a bridge between their intimate histories and the unknown future, and many opt to abandon altogether the former. Even as the nature of the Himalayan crisis is debated by outsiders, questioning whether the changes are real or not, the villagers know that in a deep way their lives will never be the same. The quick pace of modernity has little patience for the old ways.

Depending upon one's point of view, the discarding of cultural traditions may be either a necessary or a tragic thing. It may lead to advancements or to the loss of a people's capacity to survive. But if the claims of villagers are not reflected in the decisions that forever change the places where they live, then life in the Himalaya will worsen for many people. That, at least, is the fear of some, and it has a widespread precedent in the world. Under such circumstances, local communities lose a great part of their own history and abdicate control over their own life stories. The commercial changes in the Himalaya make it clear that modernity is rarely a matter of villagers seeking to meet their own needs. If that were the case, they would remain in control over their immediate worlds. It is driven, instead, by outside forces, just as the sweeping tides of history reflected the past primacies of foreign interests, and it concentrates power over resource decisions in external institutions and policies. History most assuredly repeats itself in the Himalaya, but at a quickened pace and with even more comprehensive results than before.

In the matter of livelihoods, the implications of such trends are considerable. Where native people do not direct the agency of modern life, they will inevitably become subservient to it. We find that to be already commonplace in the Himalaya; it is one of the region's foremost attributes. Meanwhile, the struggles for environmental and social justice being waged in places by local people is testimony to the determination of villagers not to let that happen. These efforts seek to retain a moral economy in the mountains that will counter the market forces emanating from the plains. One of the biggest challenges that villagers face is the shape of a new geography.

From Farm to Market

There is great interest in how to make the mountain farms more productive, especially in light of the current trends in land degradation and food shortages. Commercial interests in the Himalaya view subsistence methods as archaic and seek to enhance agriculture by replacing traditional farming with modern techniques and promoting crops for export. The idea is that the income could then be used to purchase grains, something that is probably necessary given the environmental constraints on yields and the great food deficits that prevail in many places. In Nepal, where 85 percent of the farmland is devoted to cereal crops and the total food-grain production is about 6 million metric tons, the amount of food grain produced in the country has not kept pace with population increases. Nepal consistently reports a food deficit, especially in the middle-mountains zone, where malnutrition is increasing among the low-income groups. Less than 10 percent of Nepal's mountain districts are now self-sufficient in food grains.

In many of the western Himalayan districts in India, the food situation is similarly unfortunate. The Kumaun and Garhwal regions report stagnant food-production levels but rising populations. Consequently, food is getting scarcer there, and migration from the mountains is increasing. Annual cereal production in Himachal Pradesh is about 1.5 million metric tons, twice what it was in the 1960s. But the increase there is not evenly distributed and the grains don't always reach the people who need it most. Many regions in Himachal Pradesh still need to import food from other parts of India.

The most recent estimates for food production in Bhutan, where about 95 percent of the population works in subsistence agriculture, show the country to be producing about 200,000 metric tons of food grain each year. That is less than what is needed to meet the country's food needs, prompting food imports and high levels of government subsidies in the farming sector. Nonetheless, the limited amount of arable land and the low overall yields in the countryside keep Bhutan from increasing its total food output. The country's most recent five-year plan acknowledges that self-sufficiency in food is not a realistic goal for the country.

Food deficits are also reported in the eastern Himalaya, where, despite favorable climatic conditions, the yields are lower than among the carefully terraced farms in the western region. In Sikkim, food imports make up for the shortages in local rice production. Maize, however, which is a key staple in the Sikkim diet, grows sufficiently for local needs. In the far eastern mountains of Arunachal Pradesh, the population relies heavily upon shifting cultivation, known locally as *jhum*, for its food needs. The rice yields in jhum agriculture are lower than those on the paddy terraces, but when total food production is measured, the jhum system compares favorably to agriculture elsewhere in the mountains. A major concern, though, is whether jhum can be continued under the conditions of rising populations and diminished forest areas.[1] For optimal operation, the jhum system requires extensive forested land and long periods of fallow when the fields are at rest. Neither condition is likely to exist for much longer in the region.

New technologies are being tried on the subsistence farms and commercial endeavors continue to flourish, but neither has resulted in sufficient food production in the region. That is not unexpected, since the mountains have never really been self-sufficient in food. Instead, the Himalayan cultures have survived through a combination of farming, livestock production, petty trading, foraging, and periodic migration for wage labor. That mix of activities has allowed people to persist under difficult conditions of low farm output. The government focus on grain alone ignores the full complement of economic activities that traditionally takes place in the mountains.

The new farm policies have other problems as well. They cause farmers to expand farming onto marginal lands, where problems of slope stability and soil erosion are most common. The use of fertilizers, which are necessary for the new high-yielding grain varieties, degrades the soil structure of the mountain farms and leads to the chemical pollution of soil and water. The focus on monoculture also reduces the overall diversity of the farming landscape, by adopting single crops instead of a multitude of them and by displacing many of the traditional foraging practices. As a result of these new methods, grain yields have increased on some of the well-situated, wealthier farms but have decreased on many of the marginal ones.[2] One result is the widening gap between social classes in the countryside, and another result is a continued shortage of food for many poor families.

The farm-to-market promotions of commercial enterprises exist already in different ways among the traditional systems in the Himalaya. For example, the trans-Himalayan salt trade links farms in the southern mountains with herders in the plateau region of Tibet. The agropastoral economies of the Gaddis and Gujjars in the western Himalaya and of the pastoral tribes in northern Nepal trade meat and dairy goods to neighboring communities for grains and vegetables. Local trade, based on weekly markets, has always been a distinguishing feature in the Himalaya. Not only farm and livestock products are sold in the markets, but handicrafts, clothing, soap, kerosine, oil, and other basic necessities as well. What new commercial agriculture seeks to achieve, though, is a larger scale of such activities, such that they promote national or regional wealth rather than simply meeting local needs.

The shift from subsistence to commercial farming is based mainly on the idea that specialized production geared toward the market provides the best opportunity for increasing farm income. Although that may be true under some circumstances, subsistence farming in the Himalaya is not traditionally geared toward maximizing income. Rather, it is intended to provide food security for most of the people. The measure of that security extends beyond simply the grain yields. It is located also in the social institutions that govern the production and distribution of food in the village. The farm agencies believe that by enhancing farm incomes, they will increase the purchasing power of rural people and therefore bring about greater food security, because people will then be able to purchase food during harvest failures.[3] But under the traditional systems of labor and food exchanges, cultural bonds of relationship, rather than market prices, insure survival.

Nonetheless, the majority of development plans in the Himalaya encourage the switch from subsistence to industrial farming. That usually requires more than simply a change in crops. It involves technology transfers, such as mechanization and tillage methods, credit arrangements, and the construction of agroprocessing facilities in the villages. The new high-yielding grain varieties, as well as horticultural crops, are heavily dependent on chemical additives and on the extension of irrigation facilities, making them costly for the farmers.[4] In consequence, they tend to be adopted by large farmers who rely on government subsidies in addition to their own personal investments to cover the costs.

The landscapes of the Beas and Sutlej river valleys in Himachal Pradesh, the valley of the Kali Gandaki River in Nepal, and the lower mountains in Sikkim are now covered by fruit orchards. Some of them are very big affairs. The use of fertilizers and pesticides has increased at a rapid pace in those areas. In Himachal Pradesh, for example, where temperate fruit is a booming business, the use of fertilizers has doubled in the past ten years. Fertilizer sales in Nepal between 1975 and 1980 increased tenfold in the mountains, especially in the horticulture areas, and fivefold in the lower hills where new cereal grains are grown. Reliable data is not available for Bhutan, but the level of chemical usage in the farming systems there remains quite low. Bhutan's most recent five-year plan (1992–97), however, specifically targets increased fertilizer distribution on the commercial farms.

National investments in industrial forms of agriculture provide important new technological and social arrangements in the mountains. Machinery, export markets, and investment capital, for example, lie behind the yields that result from these crop developments. Much of that production ends up going to cities in the plains, where people can better afford such types of food. One of the ironies of the new farming methods is that the food they produce is expensive while the people who need it are poor. Getting food to those who need it most remains an elusive goal in many places.

Life in the villages historically favors averting risk over investing in speculative industries. That characteristic carries people through the bad times. It requires environmental management that enables people to share common resources rather than to increase individual wealth. While villagers do participate in limited commerce, their main goal has been household and village survival. In many places, though, that may no longer be the case.[5] More people are moving into the marketplace. That shift affects more than economics; it promotes entirely new ways of living, of people relating to one another, and of cultures evaluating their natural world, and it rearranges the geography of the Himalaya.

New Regions

The princely states and colonial powers sought mainly to extract the wealth of mountain resources. That was their true imperial design, whereas the new government policies strive to bind rural people into emerging national identities. But while the means are different the goals are much the same: to gain control of the hinterlands for the purpose of territorial security and to develop the frontier economies for national wealth. Along the way, the Himalaya has taken on the affairs of the world economy.[6] We saw earlier how imperial spheres of power caused territorial conflicts between local and outside interests. The legacy of the geographical quests of the Rajput kings, the Gorkhali empire, and the colonial powers of Britain, Russia, and China all left contentious borderlands that continue to trouble the region today.

In some cases, political boundaries divide cultural territories, ignoring native claims to the land. In other instances, they are contested points of international treaties, posing problems of national security. Frequent skirmishes between India and Pakistan occur along the cease-fire line in Kashmir, established in 1947 and redrawn in 1972 but never fully adopted. The disputed Aksai Chin region northeast of Ladakh is currently occupied by China but is claimed by India. Numerous border questions linger in the eastern Himalaya along the McMahon Line, which was established during the Shimla Treaty Conference in 1914 along the crest of the peaks between India and China. Those disputes, combined with local separatist struggles and the environmental resistance movements, further complicate modern efforts toward regional and national integration.

The current efforts by Pakistan, India, Nepal, and Bhutan to expand their systems of roadways, markets, and administration into the most distant villages are meant to extend the reach of politics into these remote areas. It is hoped that they will overcome the constraints of geography, which are thought to make such places isolated, backward, and beyond the pale of modernity. Their very "backwardness," of course, makes such places appealing to the international development efforts, as well as to tourists who visit there. For the Himalayan countries, which seek to project to the rest of the world a more progressive image, such depictions may be the source of considerable discomposure. The term "to open up" (as in a territory or a country such as Nepal in the 1950s or Bhutan more recently) is common parlance in Himalayan development. It implies that such places were not exposed to the outside world prior to the efforts of western capitalism.[7] That is certainly not the case, but the extent to which regional integration occurs today does far surpass that of earlier times. Modern rural programs tackle the problems of geographical isolation first and then move toward social and technological goals.

Regional integration of the mountains follows mainly along the lines of highly centralized development. Various attempts have been made to reorganize mountain territory for the purposes of economy and conservation. Remote-area programs were popular in the 1970s, integrated rural development projects were favored by the international donors in the 1980s, and the watershed management approach was adopted in the 1990s. All these schemes exist today, so a bewildering array of regions is found in the Himalaya. In most cases, the intention of territorial programs is to link the economy, culture, and environments of the distant places with the central affairs of government.

The United Kingdom–sponsored KHARDEP (Khosi Hills Integrated Rural Development) project in the eastern Khosi Hills and the Rapti Integrated Rural Development Project, funded by the United States Agency for International Development in the western mountains are two prominent examples of the large area-development approach followed by Nepal. Smaller but similarly conceived programs, such as the German-funded GTZ (German Bilateral Assistance) program located around the town of Dhading and the Swiss development schemes in Jiri village, cover other regions of that country. Territorial concerns are central to the Agha Khan Rural Support Program located in the northern mountains of Pakistan. It administers development programs throughout the Indus mountains. Of all the Himalayan countries, Bhutan is perhaps the least involved in donor-based territorial programs. But it, too, has carved up the countryside into project territories.

Principles of conservation inform some of the new territorial approaches, especially those that follow the contours of the watersheds.

The 4,000-square-kilometer Annapurna Conservation Area Project was officially established in Nepal in 1986 to administer programs in the mountains north of Pokhara. A more recent effort in Nepal is the 2,330-square-kilometer Makalu-Barun Conservation Area, created in 1991 as a joint project of the government of Nepal and the United States–based Woodlands Mountain Institute. Those programs integrate the diverse economic needs of the villagers with national environmental and economic goals. They have become models for programs elsewhere in the Himalaya.

Although the territorial efforts utilize a regional perspective, not all places get equal attention. Geographical and social inequities occur because of the large size of the projects, the strong influence of existing centers of power, and the combative relationship that oftentimes exists between the project goals and the needs of the native cultures.[8] In Bhutan, several new parks, including the Jigme Dorji Wildlife Sanctuary, the Royal Manas National Park, and the Black Mountain Nature Reserve, are overseen by the Bhutan Trust Fund. The fund was established in 1991 to streamline conservation funding and management. Parklands now compose a total of 20 percent of Bhutan's land area. That is the highest percentage of conservation land in any Himalayan country. The parks promote sustainable economic activities, such as ecotourism and other noncompeting land uses, by enlisting the participation of local villagers in their design.[9] Bhutan hopes to prevent the sorts of conflicts between native people and parks that plague other Himalayan countries. Interventions in the Bhutan parklands include such standard initiatives as education, irrigation, and market agriculture, but their scale is kept small. A unique feature of the Bhutanese countryside are the monasteries and Buddhist forts that anchor rural settlements. These places were feudal strongholds in historic times; today they serve as administrative centers for government programs, including the conservation parks.

In much of the Himalaya, the geographical pattern of development tends to follow alongside the roadways or to cluster around established growth poles such as service centers and market towns. In the eastern sector, places such as Gangtok and Darjeeling are well connected to the national economy, but other areas in Sikkim remain poorly linked to state or regional affairs. In the western Himalaya, colonial hill stations such as Nainital and Shimla are important regional centers. The origin of the hill stations lay in Britain's colonial empire, but their modern stature is tied to India's efforts to establish greater influence over the Himalayan hinterlands. All across the mountains, public roadways, and frontier centers are key elements in the design of regions. Accessibility becomes the driving force in the modernization of the mountains. That process was greatly speeded up in the early 1990s, when the Indian government opened its roads in Spiti, Lahaul, Zanzkar, and Ladakh to public travel by issuing the new Inner Line Permits. That made many villages located in the mountainous interior more easily accessible from the plains.

Towns and Migrants

People have always been on the move in the Himalaya. Seasonal travel to pastures for livestock grazing is a historic feature of the agropastoral economy. It is characteristic of the Bukherwal, the Gaddi, and the Gujjar tribes in the Indian western mountains; of the Magar, Dolpo-pa, Gurung, and Sherpa tribes in Nepal; and of the bhotiya groups residing in Sikkim, Bhutan, and Arunachal Pradesh. In many cases, the annual distances traveled by pastoralists exceed a hundred kilometers. Such vast movements bring the herders into contact with settled farmers and mercantile centers, thus facilitating the exchange of grains, dairy products, livestock, handicrafts, and forest products.

Petty trading also has always bolstered mountain life. Long-distance trade occurs along the traditional routes between India and Tibet which traverse the Himalaya. Numerous other, smaller trading systems are located entirely within the mountains, having forged regional identities early on among the Himalayan economies and societies. To some extent, these traditional trading patterns have been displaced by new roads and markets, but many still are located in the region and define the patterns of local travel.[10]

Caravan trade is a hallmark of traditional Nepal, where porters, resting places called *chautaras,* and wayside inns still populate the rural landscape. Nowhere is this better seen than among the picturesque Thakali communities that straddle the Kali Gandaki River in the central part of the kingdom. The Thakali villages have long been the mainstay in the salt and grain trade between Tibet and India. That corridor, as well as the surrounding area, now enjoys a prosperous regional economy. Many of the old caravan routes are curious places nowadays, where yak traders, mule drivers, tourists, government officials, and merchants all intermingle freely.

The food deficits in some mountain areas have forced periodic migrations to the plains and cities for work through most of recent history. Nowadays villagers from Nepal and India even travel abroad, to such places as the Gulf States, in order to seek temporary jobs. Almost 15 percent of the population of northern Pakistan is migratory. That situation is exacerbated by the 3.5 million refugees from Afghanistan who have taken up residence in that country.[11] Ostensibly, the Afghans had come to Pakistan for a short while as victims of war, but many have opted to stay for economic or lifestyle reasons. Farther to the east, in Garhwal and Kumaun, people have been moving in significant numbers from the mountains to the plains since at least the middle of the nineteenth century, when outmigration was first reported by the British. Nowadays, with the importance of orchards and other cash-crop plantations, the mountains of India exhibit a great deal of in-migration during the harvest season.

In Nepal, where only lifetime migrants are counted by the official census takers, the main movement of people since the 1960s has been from the hills to the tarai. Lesser numbers, about 10 percent of the total migrants, move to towns, mainly to Kathmandu and the border towns, where they search for work. Almost half a million migrants from Nepal's mountain districts have settled in the tarai plains during the past two decades. Since Nepal does not bother to count its part-time migrants, the actual circulation of people in the countryside is under-reported. Overall, though, the migration trends show a mobile society in which distant places are connected in complex circulations of people and opportunities.

Himalayan migration flows also serve to connect rural places with the world at large. The actual movement of people is accompanied by the exchange of goods and services, by the sharing of information and experiences, and by the establishment of new settlements. The growing numbers of towns and cities in the mountains play an important new industrial role in the region's expanding commercial economy. Urbanization, though, remains a relatively new phenomenon in most parts of the mountains. As recently as 1981, less than one in ten Himalayan people lived in a town or city. The urban population now exceeds 20 percent. Nepal and Bhutan are still mainly rural and have less than 10 percent of their residents living in towns. But even in those countries, the importance of urban life is increasing. In 1954, for example, the Nepal census listed only ten localities with populations over 5,000. The 1991 census designated thirty-three such urban centers. During the period 1954–91, urban population in Nepal increased from a little less than 3 percent of the total population to almost 10 percent. Still, Nepal remains one of the least urbanized countries in the world, surpassed in that regard in the Himalaya only by Bhutan, which in 1990 had less than 5 percent of its population living in towns. In Sikkim, the urban population increased sharply between 1950 and 1980, from 2 percent to more than 16 percent.[12] The western Himalaya, meanwhile, contains most of the large cities in the range. Srinigar, Dehra Dun, Shimla, Mussoorie, Nainital, and several others in the region are bustling centers of population, commerce, and industry.

But overall, the small number of towns in the Himalaya reflects the persistence of subsistence agriculture, which keeps people in the villages. The low productivity of the farms also prevents the large food surplus needed to support residents of towns. The trade that has occurred in the mountains never really supported the big, flourishing towns. Some of the earliest centers were actually religious places where pilgrims congregated. Dharmsalas, not factories, were the prominent buildings in those places. In the nineteenth century, though, the British built the colonial hill stations in Srinigar, Shimla, Mussoorie, and Darjeeling. They developed into the first truly urban places in the mountains.

The new towns stem from the new regional economies that are forming in the mountains. The towns serve as industrial centers and the marketplaces for commercial agriculture. The large development projects, such as the hydropower schemes, tend to settle administrations and workers in such places, many of which have become the new frontier centers. People travel to the towns to seek work among the fledgling industries that may be located there. They work on roads, build the government offices, take factory jobs, and do all sorts of manual labor. The growth of the towns occurs foremost along the transportation corridors that connect the tarai with the mountains.

Kathmandu scene (1968).
Photo by P. P. Karan

Kathmandu scene (1997).
Photo by D. Zurick

The development of such places as Dhankuta and Dharan in the eastern region of Nepal, as Dhunche along the Trishuli River north of
Kathmandu, and as Dailekh in the far western region are examples of
such places.[13] These towns tie vast rural areas into the mountain economy vis-à-vis the network of roadways and communication arteries.
The corridors weave a fabric of new development that now drapes
across much of the mountains.

Highways into the Sky

The highest road in the world is located in the western Himalayan range. It starts in Leh in Ladakh, rises 2,000 meters in a matter of fifteen kilometers, crosses the Khardung La at 5,340 meters, and ends at the Siachen Glacier, where the armies of India and Pakistan are waging a war. The road was built for military purposes but tourists can now travel on it all the way to the Nubra Valley. Other military roads, recently opened to public travel, crisscross the mountains in Ladakh, Zanzkar, Lahaul, and Spiti. They ferry lorries laden with food and consumer goods, buses filled with villagers and tourists, pilgrims, and fancy jeeps carrying government officials and development workers into the distant mountains. The newly opened roads make these arid western mountains one of the most accessible regions in the entire Himalaya, despite the fact that deep snow and avalanches on the high passes restrict travel to only a few summer months each year.

The far-flung roads in Ladakh and Zanzkar are the most northerly extensions of a vast road system that now covers India's highlands. From the south, the roads follow the upper reaches of the valley of the Indus River by crossing from Kashmir atop the Zoji La, or by following the Beas River north to Manali and then over the Rohtang Pass into the valley of the Chandra River. On the Rohtang crossing, the Manali-Leh road in a matter of a few hours leaves behind the wet green southern slopes of the middle mountains and enters the high, arid trans-Himalayan zone. Altogether, about 15,000 kilometers of roads connect the mountain places in the state of Himachal Pradesh. The roads are vital in the developing economy of the western mountains because they allow rapid transport of the perishable fruit crop that is grown in the temperate high valleys of Baspa, Kulu, and Kangra.

Across the Indus River in northern Pakistan, the Karakoram Highway connects the Grand Trunk Road of India with the Silk Route of China. Construction on the highway began in the 1970s at a cost of $10 million and over 1,000 human lives—more than one life lost for every kilometer of road built. The Karakoram Highway now traverses some of the most rugged stretches of the Himalaya, including the magnificent Hunza Valley, before it crosses the 4,880-meter Khunjerab Pass and enters China at a place called Sust in the Xinjiang province.[14] It may well be the most spectacular road in the entire world. Clinging to rocky precipices, crossing unstable glacial outwash, undercut by torrential streams and rivers, the Karakoram Highway is an engineering marvel. Although few exports currently leave the mountains, other than apricots from Hunza and timber felled in the Darel Valley, it is reckoned that the road will play a key role in the future economy of the upper Indus valley. The Karakoram Highway is already indispensable in delivering food supplies to the northern valleys.

In Nepal, which had only 289 kilometers of passable roads in 1964, the flurry of road building continues apace with the kingdom's quest for modernity. The country now has over 2,500 kilometers of paved roads, almost all of which are located in the lower mountains and in the tarai plains. The number of and miles of high mountain roads increase every year. Dynamite blasts continually rock the countryside along the major river valleys. Pokhara is connected to Kathmandu, with feeder roads heading north to Gorkha. A new road leads north from the Kathmandu Valley to the Langtang Valley via the Trishuli River and the vil-

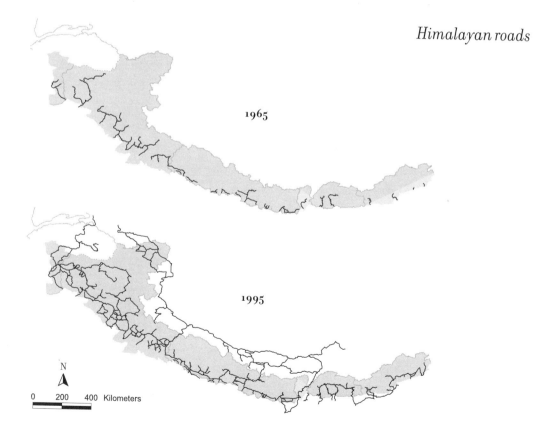

Himalayan roads

1965

1995

N

0   200   400 Kilometers

lage of Dhunche. Other roads connect the valley with the Tibet border at Kodari and with the village of Jiri below Gauri Shankar along the Mount Everest route. Still more roads lead south from the valley to the Indian border.

In the far western region of Nepal, a road now leads into the highlands north of Birendranagar; it is planned to eventually reach the isolated town of Jumla. Another road is scheduled for the eastern region, to connect the village of Salleri in the Solu-Khumbu region with the tarai border. But the most ambitious road-building project is the one promised to the people of remote Mustang. That road would follow the Kali Gandaki River for 150 kilometers to the village of Lo Manthang. It has a price tag of $32 million U.S. The engineering requirements of the Kali Gandaki road would be formidable indeed; it would traverse the deepest gorge in the world and some of the harshest terrain in the entire Himalaya.

The construction of new roadways constitutes a major component of development in the eastern Himalaya as well. Sikkim now has about 1,000 kilometers of motorable road, while Bhutan's road network exceeds 2,000 kilometers. Roads are still scarce in Arunachal Pradesh, but the Indian government has made them a top priority in that region. The fact that roads occupy such prominence in the development plans of all the Himalayan countries reflects the fact that they are crucial for the kind of economic advancements proposed in the mountains. The roads in themselves do not produce economic growth, but they make it possible for more commercial development and resource extraction to take place.

The roads, however, do more than facilitate economies. They draw a new geography. The corridors create linear landscapes, new edges, in the mountains. Migrants follow the narrow alignments in the hope of securing construction employment, shaping the settlements of shantytowns along the roadways. Tourists and pilgrims travel along the roads amid the promotions of hotels and restaurants. Billboards and flashy advertisements for tourism services are now commonplace along the well-traveled routes. In formerly isolated localities, the introduction of a road forever changes the nature of the place.[15]

The cost of building the Himalayan roads is high, in terms both of money and of human lives. Scores of workers are killed when roads are built. Even more deaths take place when workers repair the roads af-

Much of the road building in Nepal still depends on hand labor for breaking rock and clearing slopes. Photo by P. P. Karan

ter avalanche damage and mudslides. The working conditions are treacherous. In September 1995, more than fifty laborers were buried when a portion of the Kulu Valley highway slid into the Beas River. But the hazards of roadwork do not deter the legions of migrant workers who come from across the Himalaya to labor on the roads; they see the possibilities of employment on the road crews as being one of their greatest hopes. The huge initial costs of the road constructions are usually financed by international aid, but the recurring expense of maintaining the roads must be borne by the Himalayan countries. Such costs may be staggering. In Nepal, for example, the environmental damage to mountain roads between 1980 and 1993 required an outlay of over 2.5 billion rupees for the repair work alone. Furthermore, the environmental degradation caused by the roads may be great. Estimates show that the soil loss from roads in Sikkim and Garhwal amounts to 2,500 tons for every square kilometer of roadway. In Nepal, up to 9,000 cubic meters of landslides occur per kilometer of mountain roads during construction. Overall in the Himalaya, poorly constructed roads cause the loss of up to 150 tons of soil per hectare per year.[16] By contrast, soil loss from the degraded Himalayan forests is estimated to be less than three tons per hectare per year. The Himalayan roads, built for economic advancement, may irreparably damage the mountain landscape and further degrade the conditions of rural life that they are meant to improve.

The Idea of Sustainable Development

The Himalayan roads support a type of economic growth in the mountains that is based upon commercial exchanges and the export of natural resources. According to many planners, the damage done by the roads is offset by the economic advantages they bring to the villages. That is a persuasive argument in the rural areas. Where roads do not exist, they are sorely missed. Indeed, most villagers, when asked what they would most like to see come into their lives, will answer roads and only afterward mention other tangible items of development such as electricity.

The roads, dams, and other infrastructures are the products of national initiatives. They are highly visible and have valuable political currency. But their impacts on society and the land are great and not always understood. The adverse human and ecological consequences of such programs may actually advise against them. Some Himalayan planners now call for alternative economic programs, ones that are smaller in scale and promote greater social equality. The term "sustainable development" is given to such efforts because the planners advocate restoring the natural environment while at the same time combating poverty.

The sustainable-development model remains elusive in the Himalaya because it lacks political backing. The big development projects still corner the market. But the precepts of sustainable development guide many nongovernmental organizations working for change in the region. The proceedings of a conference held in 1994 at the International Center for Integrated Mountain Development (ICIMOD) in Kathmandu, under the title "Sustainable Development of Fragile Mountain Areas of Asia," showed the extent to which sustainable development has become an accepted model for mountain development among the more progressive development agencies.

According to the ICIMOD initiatives, sustainable development should be locally determined and of a comprehensive design. Its primary focus should be on the attributes of villages, whereby the assessment of local circumstances outweighs those of purely national interests. Sustainable development is at odds with many of the contemporary trends, especially those marked by the roads, reservoirs, resource extractions, and other large infrastructural advancements.

Those things still consume much of the region's development budgets, but according to the ICIMOD conference in Kathmandu, the focus of sustainable development should be on the mountains. Unfortunately, the plains are still the center of attention for many regional politicians. In an opening address to the Kathmandu conference, Ms. Savitri Kunadi, Vice-Chairperson of the Commission for Sustainable Development, requested that ICIMOD "sensitize the international community towards this complex interlinkage between the activities in the plains which may be beneficial for the sustainable development of mountains, the effects of conservation of the mountain ecosystem on the plains . . . and similarly how the adverse activities of one have substantial and significant detrimental effects on the other."[17] Her statement is significant because it highlights the fact that, despite the considerable intellectual and activist efforts to disengage the mountains from the plains, the two regions remain tightly wedded developmentally in both practice and policy, with the outcome still in favor of the plains.

Logging exacts a heavy toll on the forests located in the Nepalese tarai south of Hetauda.
Photo by P. P. Karan

The Control of Nature

WIDESPREAD POVERTY remains the most confounding and paradoxical dilemma in the Himalaya. On the one hand, it is both the result and cause of land degradation in the mountains; on the other hand, it compels much of the development that results in additional ecological damage. That duality exists because, despite the rhetoric of sustainable development or the promise of conservation, population pressures continue to beset the land while at the same time the course of change in the mountains still follows the way of big development. The most widespread transformations in the Himalayan landscapes result from such innovations as commercial agriculture, water-diversion schemes, infrastructures, and mining operations and other forms of commercial resource extraction.

These activities may generate considerable wealth, but it is not

13

evenly distributed among the mountain people. The Himalayan communities continue to be some of the poorest places on earth, while their resources continue to disappear at fast rates. Commercial resource extractions do little to restore the health of the Himalayan environment, leaving that to conservation groups, as if the two goals were indeed separate concerns. Where alternatives are imagined, such as with some of the Himalayan parklands or with the grassroots projects, they reflect the determination of local communities. For that reason, the advocates of more responsible forms of development call for the full partnership of villagers in the design and management of economic programs. In their view, the uniqueness of the mountain world, its dynamic qualities, its environmental fragility and diversity, as well as the resilience of its cultures, should all occupy a central place in development designs.

But the Himalaya still is being depleted of its wealth by the short-term plans of business interests. They are the most visible forces of change in the mountain landscapes and show that, in the dialogue between sustainable development and big development, the former remains rhetoric while the latter shapes the lay of the land. Foremost among the big development interventions are those that exploit timber and agriculture products for the marketplace, divert water for energy generation, and dig into the deposits of the Tethys Sea for their potential mineral wealth.

Forest Wealth

The loss of forests has been the primary concern of environmentalists in the Himalaya; it is linked to a wide set of ecological disturbances that contribute overall to high rates of land degradation and diminished natural productivity. Today, less than a quarter of the range remains under forest cover. Each year, over 126,000 additional hectares of forest are removed. The annual rate of forest loss varies from 0.2 percent on average in the western mountains of India to over 1 percent in neighboring Pakistan. Much larger averages are recorded at the district and community levels, so it is hard to figure out what is really going on.

In most assessments of forest degradation in the Himalaya, the main concern is still with the growing numbers of humans and farm livestock, which are increasing at annual rates of more than 2 percent (humans) and 3 percent (livestock). People need food and fuel; livestock eat fodder. But local pressures on the forests are not restricted to the basic needs of farmers. The villages also employ small-scale industries that require forest products. House building and woodworking; charcoal making; metalsmithing; ghee, cheese, and alcohol production; fruit processing; paper making; resin tapping; and medicinal herb collecting all take a toll on the forests. The promotion of such commercial industries is necessary for the village economy, especially where agricultural yields are low. Hence, they receive the support of both government economists and local activists, though they do have an impact on the forests.

The cutting of the Himalayan forests by commercial timber merchants is also commonplace in the mountains. The commercial operations are often facilitated by ineffectual forestry policies or by the collusion of low-paid forest guards and lumbermen. The combined impact of village industries and the export timber concessionaires surpasses in many places local subsistence pressures, but the villagers still get most of the blame for deforestation.

In Nepal, where the forest conditions are considered to be some of the worst in the region, their degradation is considered to be linked almost entirely to the subsistence economy. In some mountain districts, where the population pressure is especially great or where community forest control is very weak, the degraded forest cover may indeed be explained by fuel-wood and fodder collection, but the overall change in mountain forests during the past decades is minor—about 0.3 percent decrease per year.[1] However, the lowland Nepal forests have been depleted by more than a quarter of their area in the past two decades. The current rate of forest loss in the foothills is almost 4 percent a year. In part, that is due to immigration and farmland clearing, but to a great extent the change results from the activities of the commercial timber industry.

The marketing of Nepal's trees is organized by the Timber Corporation, the Fuelwood Corporation, and the Forest Products Development Board. Their interests overlap and belie the limited conservation efforts of the Ministry of Forests. The Fuelwood Corporation was established in 1962 to ensure a steady supply of fuel wood for Kathmandu. It now supplies, at government-subsidized prices, more than half the firewood used in the Kathmandu Valley. Most of the lorries enter the valley from the south, where the tarai is located, and head directly for the brick factories, where the bulk of the fuel wood is burned to make building materials.

In the Indian Himalaya, where the commercial exploitation of forests has a much longer history, revenue from forest products continues to increase. In Garhwal and Kumaun, for example, forest sales—mostly the sale of timber logs—jumped by 25 percent in the past decade. In Himachal Pradesh the sale of timber from private trees has been banned officially since 1982, but the State Forest Corporation reported timber sales in 1993 of 0.5 billion rupees. The overall revenue that the Forest Department gained by timber royalties and taxes grew from 25 million rupees in 1975 to over 860 million rupees in 1993, an increase of well over 3,000 percent.[2] Even such high figures are conservative. By its own admission, the Forest Department reckons that only a quarter of the forest area in the state in Himachal Pradesh is actually demarcated and surveyed, so the profits from the forests that enter into the official records are much less than the full wealth that accrues from the forest encroachments, illegal timber sales, resin tapping, and extraction of minor forest products.

In Bhutan, where forests still cover more than 60 percent of the land area, timber sales in 1988 exceeded $11 million U.S. and now account for almost 10 percent of total government revenue.[3] Bhutan timber products are exported exclusively to India and Bangladesh. The Department of Forest in Bhutan was established in 1952 to promote revenues from the national forests, but now it has mainly a conservation purpose. In order to discourage the illegal cutting of its forests, Bhutan has nationalized all logging in the kingdom. The overall low rate of forest loss in Bhutan shows how geography, as well as government policy, can limit the scope of commercial timber cutting. Most of Bhutan's prime forest land remains inaccessible to the loggers. However, as new roads are built, there is concern that commercial forest activities will expand into the more remote places.

TREE HUGGING

In the western Himalayan regions of Doon Valley and the Garhwal hills, where timber contractors have been especially busy, the level of community resistance to forest cutting is strong and well organized.

People's lives are jeopardized there by commercial timber operations. The most famous of this area's environmental struggles is the Chipko Protest, which is based on the Gandhian ideal of nonviolent resistance. Villagers have led symbolic marches, have held "tree hugging" demonstrations where people wrap arms around the trunks of trees, have done scientific research on the forest and started reforestation programs and related conservation policies. The protest led directly to a moratorium on live-tree cutting above 1,000 meters in a 40,000-square-kilometer area of Garhwal and neighboring Kumaun.

Offering more than a simple conflict over resources, the Chipko Protest embraces an entire philosophy. In the commercialization of individual tree species, the Chipko leaders assert, forest ecosystems are reduced to merely economic appraisals, when for the villagers the forests are the very sustenance of life. The struggle waged by the Chipko activists is not about the specific endeavors of individual timber contractors, as detrimental as those may be for the region, but about a whole worldview that equates nature with economic measurements. According to Sunderlal Bahuguna, an early Chipko leader from Garhwal, "Protection of the forests is not an isolated problem. It is closely connected with the process of development. In a society where affluence is the goal of development, all natural resources including the forests are eventually exhausted . . . Such protection will, on one hand, ensure the coexistence of trees and human beings, and on the other hand, build the foundations of a decentralized society in which the basic needs of humans and other beings are fulfilled from their surroundings."[4]

In recent years, and for a variety of reasons, the influence of the Chipko Protest has lessened. It lost its effectiveness once it took on a largely symbolic role. Politics proved, in the end, to be its downfall. Nonetheless, Chipko remains an important model for community environmental resistance around the world.

In the Doon Valley, located north of the town of Dehra Dun, community activists joined together in the early 1980s to form the Friends of Doon Association. Their intention was to slow down commercial exploitation of the valley's forests, which have declined by a quarter in the past two decades. Ninety-eight percent of all the wood extracted from the valley goes to industry, leaving little for villagers. The scarcity of fuel wood has caused farmers there to burn cattle dung instead of putting it on their fields for fertilizer. The Friends of Doon, like the Chipko protesters and participants in other environmental struggles in the mountains, strongly advocate local authority over the use of natural resources. That does not preclude commercial forest development, but demands that it should be of a manageable size and under the authority of villagers.

A Case of Apples

The orchards have become an important new feature in the agricultural landscapes of the Himalaya. At the lower elevations, subtropical fruits are common. In the temperate regions, apples and peaches are grown. In places where land is unsuitable for cereal grain production, this new horticulture is a promising economic innovation, and its potential to generate revenue is now recognized by all the Himalayan countries. Nowhere is that more apparent than in the apple-growing areas of Kashmir, Himachal Pradesh, and north central Nepal. The familiar European apples—Jonathans, Macintosh, and Delicious are all planted in these localities with considerable success. The orchards once were an incongruous element of the landscape; now they are ubiquitous in the western range.

The cool climate of the mountains gives them a comparative advantage over the plains in raising temperate fruit. The lowland cities, meanwhile, constitute large markets for the produce. Where roads exist to transport the fruit, extensive land areas are now devoted to orchards. In the warmer climates of Sikkim and southern Bhutan, citrus fruits and cardamom are grown. Much of the land in Darjeeling is used for tea gardens. Such horticultural plantations have transformed the mountain landscapes across wide areas.

In Nepal, government investment in orchards started in earnest in the late 1960s, when the country established a number of fruit and vegetable programs. About thirty government farms have since been developed which provide nearby farmers with planting material and technical assistance. Most orchards in Nepal, though, remain small household affairs. The more recent establishment of large apple orchards in the Khumbu region and along the upper Kali Gandaki River by the Thakali villagers indicates the potential for expanding commercial fruit production. The orchardists not only grow fresh apples

An apple orchard in the village of Kotgarh, Himachal Pradesh, India. Photo by D. Zurick

and apricots but dry them, turn them into preserves, and distill brandy from them. The area under fruits in Nepal increased from 32,000 hectares in the mid-1970s to 56,000 hectares by the beginning of the 1990s. The target for the new millennium is 87,000 hectares of orchard.[5] But the lack of roads and the problems of marketing the perishable fresh fruit remain big obstacles in the development of Nepal's horticulture industry.

In the northwestern Himalaya, horticulture is already big business. About 500,000 hectares of land are devoted there to commercial fruit and nut production. Forty percent of the orchards grow apples, which account for over 80 percent of the total fruit crop grown in the mountains. Apple production has increased tenfold over the past decade alone, with major impacts on the regional economy, village society, and the environment. The most extensive areas of apple production are in the mountains above the Kashmir Valley and in the large river valleys in Himachal Pradesh. The upper Beas River area and the valleys of the Sutlej and the Baspa are all important growing areas.

Apple growing dates back to the 1930s in Himachal Pradesh, when Reverend Stokes, an American missionary, imported apple seedlings

A government agroprocessing facility in the apple-growing region of the Sutlej Valley, India. The spin-off industries of horticulture provide new jobs in the mountains. Photo by D. Zurick

from North America and planted his orchard in the village of Kotgarh, overlooking the Sutlej River. The original Stokes orchard, situated near Highway 22 leading from Shimla to Rampur, remains a prominent feature in the local landscape. A second orchard was established at the same time in the nearby valley of the Beas River by Captain Lee of the British army. Both these early orchards were the hobbies of landed gentry from the West, who chose to live for a time in the Himalaya, but their apple trees proved to be the beginnings of a rapidly growing economy. By 1950, 400 hectares of land around the original valley estates of Kotgarh and the Beas River were planted in apples. By 1965 the apple orchards had increased to over 12,000 hectares, and by 1993 approximately 70,000 hectares of land in Himachal Pradesh were devoted to apple production. Other temperate-climate fruits, such as peaches and pears, as well as nuts, occupy an additional 45,000 hectares of land in the state. The gross value of apples in Himachal Pradesh was 3 billion rupees in 1994. Clearly, they have been a major success for the region.

The high level of government investment in fruit orchards reflects their potential to generate revenue for the state. The investment in fruit production by the Himachal Pradesh Department of Horticulture increased from 28 million rupees in 1970 to 185 million rupees in 1995. Much of that has gone toward providing credit, extension services, and technology, as well as building fruit-processing and storage facilities in the villages. The large orchardists are influential leaders in the state. Power is one result of their newfound wealth.

The tourism industry in Kulu, which coincides with the harvest season, provides a strong local market for the apples grown in the valley of the Beas River. Elsewhere, orchards stretch for many miles along the Sutlej River, from Kotgarh village to the Baspa valley and beyond to Spiti, where they are found at elevations of more than 3,000 meters near the village of Tabo. There, the orchards, watered by irrigation canals, create green splashes in the stark, arid landscape and brighten the autumn landscape with their red fruit. Orchards are found in amazing places along the Spiti River, clinging to steep canyon walls and cascading down the tributary valleys.

Total apple production in Himachal Pradesh has increased at a pace that matches its expanding area, from 120,000 tons in the 1970s to over 350,000 tons by the mid-1990s.[6] The economic role of apples is derived not only from the actual fruit but also from the orchard nurseries, packing stations, warehouses, and fruit-processing plants. These industries straddle the main roads into the apple-growing mountains. In response to high demand, the government has established 300 centers of horticulture research and training to support the orchards. All that activity gives a busy, industrial look to the mountains.

Apples and the Environment

Where they make good use of steep slopes and occupy unique ecological niches, the orchards are an effective land use. Under proper management, their trees help to maintain slope stability. Their roots bind soil together. The fallen leaves and debris provide mulch. Their canopies lessen the impact of rainfall on soil erosion. But the orchards also introduce a monoculture into the complex farming systems, thereby diminishing the diversity of plants grown and simplifying the food base of the villages.

The greatest conflict posed by the orchards is competition with

Apple orchards cling to the very steep slopes above the Spiti River, India. Photo by D. Zurick

food farms for land. Since apples are primarily grown for export to the plains—fully 98 percent of the production is shipped out of the region—their role in local diets is negligible. Studies in the village of Kotgarh, where the orchards are well established and occupy 75 percent of the good farmland, show that crop intensities have increased on the land that still grows food. The steep slopes in the region cannot support the more intensive land uses, and as a result soil erosion has become a recent problem for many farmers. Cereal grains have also become more scarce in Kotgarh, since land is taken up by apple trees.

The productivity of the apple orchards in Kotgarh depends on water and chemical inputs. Field irrigation is necessary to offset moisture stress in the trees during the dry season. Fertilizers control yield, color, sugar content, and acidity of the fruit. The amount of fertilizer used in Himachal Pradesh exceeded 40,000 metric tons in the mid-1990s, most of it applied to the orchards. The government heavily subsidizes fertilizer sales; the total transportation costs associated with fertilizer use is paid for by state taxes. In addition to the fertilizers, the orchards use large amounts of pesticides, which must be applied to the trees seven times a year to combat various insects and diseases. The distribution of pesticides, 250 tons in 1994, is handled by the state Agro Industries Corporation, a quasi-government operation.

The Kotgarh orchardists pride themselves on being known throughout the region as progressive farmers. They use an array of modern technologies and chemicals in their orchards. Their apple yields are on average three times higher than those of farmers elsewhere. But the success of the Kotgarh orchards, and of other nearby apple-growing regions, has environmental costs. Chemical loads in the streams and rivers have risen sharply in the past decade as a result of fertilizer and pesticide runoff. Comprehensive studies on water pollution have not yet been done, but several environmental agencies plan to study the Sutlej Valley.

One of the most serious environmental impacts of the apple industry is created by the use of trees for packing crates. The apple boxes, which have a capacity of eighteen kilograms of fruit, are made of fir, pine, and poplar woods. More than 12 million crates are built each year to ship the apples out of Himachal Pradesh. The wood used in constructing the apple boxes equals 50,000 average-size fir trees per year. In Kotgarh, the entire supply of wood for the crates is obtained in the local government forests. In that way, the villagers end up subsidizing the apple industry. By contrast, most of the fuel wood and fodder needs of the villagers are met by their own private woodlands.

Two hundred sawmills operate in the Kotgarh region alone to provide wood for the apple growers. The amount of timber cut from the government forests in Kotgarh just for apple crates increased fivefold between the 1960s and the 1990s, from less than 5,000 cubic meters to more than 20,000 cubic meters.[7] To lessen the timber requirements of apple crates amid the expanding orchards, wood substitutions are being explored by some growers. Corrugated fiberboard boxes have become common in some places. They use a third of the wood required by the old crates. Plastic containers are also being considered as a substitute for wood.

The apple orchards have increased wealth in the mountains. That is clear. The distribution of that money, of course, is highly skewed. Kotgarh, for example, is a prosperous area compared to its surrounding villages. That sets it apart, even in the apple-growing region. But the wealth that accrues in Kotgarh has produced new economic strata in the community. Study of landholdings, household income, and consumption levels shows widening gaps between the orchardists and the other farmers. The initial capital outlays and the ongoing maintenance costs of the orchards keep poor farmers from investing in horticulture. Meanwhile, the government subsidies are directed toward the wealthiest fruit growers. The income from production taxed by the state has yet to go into comprehensive programs that provide services or alleviate poverty among the poorer households.

On the one hand, apples do well on steep mountainsides, which otherwise could not sustain agriculture. In those places, the apple orchards are very good uses of the land. Where they are carefully integrated into the farming systems, which include both subsistence grain farms and commercial enterprises, fruit orchards are positive additions to the temperate Himalayan landscapes. But in some places, for example in the Baspa valley and along the upper stretches of the Spiti River, conflicts arise as the orchards expand onto food-growing farmland. In these cases, overall food availability may actually decline, as it has in Kotgarh.

Because of the heavy demand for land for apple orchards, the Kotgarh farmers do not produce sufficient food to meet the local needs. The small farmers report the largest food deficits, more than three times greater than those reported by the medium and large farmers.[8] The shortages must be met by purchasing food, thereby placing a premium on cash employment. In the new economy of horticulture, the orchards are viewed favorably by many villagers because of the jobs they may provide. Almost half the villagers in Kotgarh work in the apple business, either full time in the processing plants or as casual laborers in the orchards. Additional people come from throughout the Himalaya, as well as from the plains, to work in the orchards during the apple harvest. Full-time farming is giving way to part-time wages.

Some orchard tasks, such as pruning and field maintenance, do employ workers the year round. Most of the labor, though, is seasonal, and the greatest opportunities exist during the autumn months, when teams of workers, organized by fruit contractors, arrive in the villages to pick and package the fruit. In places such as the valley of the Baspa, they constitute a sizable migrant population. The ancillary work of food-processing, sawmills, and warehouses provides additional la-

Migrant workers in the valley of the Baspa collect the autumn apple harvest and prepare it for shipment to the lowland plains. Photo by D. Zurick

bor opportunities. Most of the transportation and marketing work, though, is done by entrepreneurs based in the lowland cities. They work in advance to secure harvest contracts with the growers and arrange their own pool of contract workers.[9] Despite the problems, the general perception is that the mountain economies benefit greatly from apples; that insures their continued success, despite the adverse social and environmental costs.

Vanishing Rivers

In the view of many people, the greatest potential for national wealth in the Himalaya lies in harnessing the mountain rivers for irrigation and hydropower. The development plans of all the mountain countries contain schemes to expand energy output for domestic purposes and to export electricity to large cities located in the plains. The schemes range from the micro hydroelectric works in the remote villages of highland Nepal and northern Pakistan, which generate less than thirty kilowatts of power and are intended solely for village electrification, to some of the biggest dams in the world. The latter include the controversial Tehri Dam under construction in Garhwal. When it is completed, it will have a capacity of 1,000 megawatts, enough power to light up all of India.

The potential of the large dams to advance the economic growth of the Himalayan countries is considered to be truly enormous. It is thought that they will produce the power needed to drive the engines of industrial expansion throughout the mountains, with a good deal left over to be sold to the plains for cash. For government officers making decisions about life in the mountains, the economic prospects of hydropower development far outweigh its adverse consequences. In addition to the energy products, the reservoirs will trap water for agricultural uses. The desire to increase food production is based upon necessity, and irrigation is the one technology that can help to achieve it. For such compelling reasons, all the mountain countries except Bhutan consider building the massive dams along the Himalayan rivers to be vital for their future economic and social development.

But the destructive impacts of the large dams loom large on the jagged mountain horizon. An important international conference was held in 1994 in Kathmandu to consider sustainable development in the Himalaya. There was no mention at that conference of the need for large dams.[10] Instead, the conference participants, many of whom were energy experts, proposed that the energy requirements of the Himalaya could be obtained by a diverse blend of renewable sources, including biomass and solar energy as well as micro hydropower projects. The small hydroelectric schemes, with capacities under 200 kilowatts, are thought to be most affordable at the village level. They also minimize environmental damage. Their modest size will not generate huge regional wealth, but they will meet the needs of local communities. The large dams would change the entire hydrology of the Himalayan range as they staunched the flow of rivers, inundated thousands of kilometers of land, and displaced entire villages. Moreover, the life spans of the large dams will be shortened by the heavy sediment loads of the mountain rivers they seek to trap. That diminishes their overall economic performance. These problems are widely acknowledged, but they have yet to dampen the enthusiasm of the big-dam proponents.

The Tarbala Dam was built on the Indus River in Pakistan with the future in mind. It is one of the world's largest earthen dams. The power output currently provides a third of the electricity used in Pakistan. But the reservoir is filling up fast with sediments, so that it may have only a few more years left. Other Himalayan reservoirs face a similar fate. The high rates of geological uplift and erosion cause a steady stream of sediments to flow into them. When measured against the massive landscape changes they produce, the longevity of the dams should be a critical concern. Too often, it is not. Perhaps the most important issue, though, is the active seismicity of the Himalaya, especially along the front range of the mountains where most of the dams are built. It threatens the very structures of the dams and the lives of those who live near them. If the dams collapsed in an earthquake, the downstream effects would be truly devastating.

In light of these elements, the dams play a contentious role in the future of the Himalaya. In a grand way, the water diversions and hydroelectric schemes realign the entire flow of water and lay of the land. They should be considered, therefore, in regard to the full complement of water issues in the mountains and not just as engineering problems. Already, the water resources in the Himalaya are heavily taxed: threatened by destroyed watersheds, by depleted groundwater, and by pollution. The dams and river diversions seek to further diminish the natural flow of water in the mountains.

THE FLOW OF WATER

The surface runoff in the Himalaya is determined by the distribution of rainfall, the locations of snowfields and glaciers, and the course of the rivers themselves. The monsoon is instrumental in the formation of regional climates; its role also is great in determining the flow of surface water in the mountains. The two great rivers of Indus and Brahmaputra wrap around the Himalaya from west to east in a vast, watery embrace. A third major river system, the Ganges and its tributaries, drains much of the southern face of the mountains. Some of the mightiest rivers of the Ganges basin, the Karnali, Gandaki, and Arun for example, have their headwaters on the Tibetan plateau and cut through the heart of the mountains. Others originate among the Great Himalaya peaks, from the snows and glaciers that accumulate there and course their way south through the middle mountains. In all, an estimated two-thirds of the combined flow of the Ganges-Brahmaputra rivers, where they join to empty into the Bay of Bengal, originates in the Himalaya and the outlying plains.[11]

The Sutlej River near its headwaters along the India-Tibet border. Photo by D. Zurick

The Indus River flows west from Mount Kailas before turning south at Nanga Parbat toward the Arabian Sea. The annual flow of the upper Indus, before it is trapped by the Tarbala Dam, is an estimated 115 billion cubic meters. Most of the discharge entering the upper Indus River comes from the Karakoram Mountains, by way of the Gilgit, Hunza, and Shyok rivers, as well as numerous smaller tributaries. But the upper stretches of the Indus also drain sections of the northern slopes of the Himalaya, especially where it traverses the arid highlands of Ladakh. In all, it represents a formidable hydrologic boundary in the western section of the range.

Farther downstream, the Indus gets its water from the rivers that drain the south-facing slopes of Kashmir and Himachal Pradesh. It is possible to stand at the divide of the Kun Zum La, between the high valleys of Spiti and Lahaul, and watch as the streams flow both east and west. The latter streams descend to the Chandra River and eventually enter the Indus system. The former lead to the Spiti River and ultimately add to the river flow of the Sutlej basin. At the southern end of the Spiti river, looming above the magnificent valley of the Baspa River, is the imposing face of Kinnaur Kailash Mountain. The meltwaters from the glaciers and snowfields located on its northern face drain into the Indus and eventually into the Arabian Sea; from the southern exposure, the waterflows merge with the Ganges River to flow into the Bay of Bengal. Standing atop that great watershed divide, the grand shape of the Himalaya assumes a tangible reality.

Much of the water flowing south from across the Himalaya is captured by the Ganges River, whose headwaters are located in an ice cave near Gangotri, below Kinnaur Kailash. The rivers of the western Himalaya add about 50 billion cubic meters of water to the Ganges River. An additional 145 billion cubic meters are provided by the watersheds of Nepal. Altogether, the Himalayan rivers compose more than 40 percent of the Ganges before it meets the Brahmaputra. The remainder of the flow comes from the Indian plains, where heavy monsoon rains feed the lowland tributary rivers.

The Brahmaputra River—or Tsangpo, as it is called in Tibet—also originates in the Kailas range near Lake Mansorawar. Where the Brah-

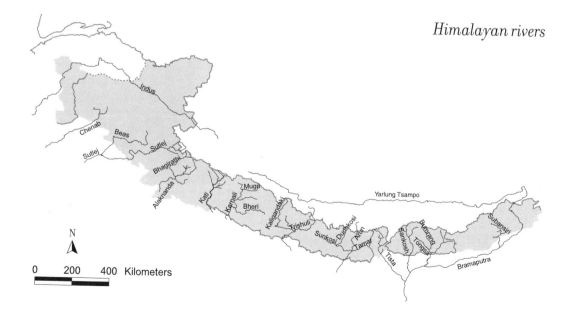

Himalayan rivers

maputra empties into the Assam Valley near Pasighat, after spanning almost the entire length of the mountains, the river contains an annual discharge of 200 billion cubic meters. Seventy percent of that volume comes from the Tibetan plateau, while the rest is runoff from the mountains of Sikkim, Bhutan, and Arunachal Pradesh.[12] Considering the volume of water it contains and the landscape that it drains, the Indus-Ganges-Brahmaputra river system is of a world class.

The water resources of the Himalaya have had a great historical importance for the people of South Asia. Many of the mountain rivers, especially the Ganges and its tributaries, are revered as deities. Numerous spots along the rivers, especially where they meet, are holy places. Pilgrims go there to wash away their sins. The rivers thus have a spiritual meaning for the native people of the mountains. But the rivers are also life giving in a practical way. Villagers divert them to water their fields and grow food. The traditional irrigation systems are one of the marvels of the indigenous landscape.

The scale of the modern water diversions is much greater than that of any of the traditional practices. These new practices lie beyond the pale of religious sentiment, grounded instead in the logic of Western science and economy. The divergence between the traditional views and uses of water and those of modern society has caused conflicts in places such as Rathong Chu in Sikkim, where native practices give the flow of rivers a deeply religious importance and where the planners seek to put a huge dam. An entire world will be flooded at the event. Even Western scientists, informed about the seismic hazards, caution against the big dams in the Himalaya. Their voices are lost, though, in the din of the turbines and rushing pipelines.

Power and Dams

It was the British who first thought about damming the Himalayan rivers. As early as 1908 they surveyed the Sutlej River with that in mind. But only with the independence of India in 1947 were serious government plans made to harness the energy potential of the mountain water. Jawaharlal Nehru, India's first prime minister, remarked, "I often think that not only is this great mountain chain a boundary and a frontier of India rising like a great sentinel, inspiring so much of our culture and thought throughout the ages but that this mighty chain is

The Beas-Sutlej Link Project diverts water between the two Himalayan rivers in order to generate electricity for export to the Indian plains. The project is one of the largest hydropower schemes in northern India. Photo by D. Zurick

Hydropower development along the Sutlej River has created an industrial corridor through the mountains. Photo by D. Zurick

also an untapped source of energy. The energy flows out in great rivers. If only we could utilize this mighty reservoir of energy to full purpose."[13] His wishes came true in 1963, when the 226-meter-high Bhakra Dam was built on the Sutlej River near the town of Bilaspur. It was designed to generate 1,200 megawatts of electricity for the plains cities and to irrigate 3.5 million hectares of land in Punjab and Rajasthan.

In 1974, a second dam was built at Pong on the Beas River, producing 360 megawatts of energy. More recently, the massive Beas-

Sutlej Link was created to divert water from the Beas through a forty-eight-kilometer pipeline into the Sutlej, thereby driving turbines to the tune of 660 megawatts of electricity at the Dehar power plant. Nine additional run-of-the-river hydro schemes are situated along the upper Sutlej River in the district of Kinnaur. These schemes do not require massive reservoirs, but the pipelines needed to move water through the turbines have necessitated massive excavations of the mountainside. The run-of-the-river schemes, which the government planners insist do no environmental damage, have transformed the Sutlej valley into a virtual industrial landscape. Dusty roads, sprawling shantytowns, brothels, hotels, power lines, and factories now cover much of the river's course.

The most controversial hydro scheme in the western range is the Tehri Dam, located on the Bhagirathi River near Garhwal.[14] The 260-meter dam project was approved by the Indian Planning Commission in 1972. Once completed, the Tehri Dam will provide 1,000 megawatts of power and irrigate 270,000 hectares of land in the Uttar Pradesh plains. The main problem with the Tehri project is that the reservoir will displace over 100,000 people. A total of twenty-three villages and 5,200 hectares of farmland and forests in the mountains will also be flooded by the impounded water. The town of Tehri, itself, an important cultural and religious center, will be fully submerged.

Local opposition to the Tehri Dam began in 1977, when villagers organized into a community resistance group called the Anti-Tehri Dam Committee. Its concerns were centered first on the displacement of people living upstream of the dam. A later issue was the safety of the dam itself. The community group enlisted scientists to study the seismic hazards of the project. The dam was to be situated close to the Srinigar thrust fault, where earthquakes are likely to occur. The technological hazards of the Tehri Dam are now well documented, and show that there is a major safety issue. The Anti-Tehri Dam Committee also looked into the economic projections of the dam and found them to be unrealistic. The high sediment rates of the Bhagirathi River will clog the reservoir with silt, shortening its life span and diminishing the long-term benefits of the project. Considering the safety risks, the displacement of villagers, and the expense of the dam, the Working Group on Environmental Impact of Tehri Dam concluded that it should not be built.[15] Despite those objections, construction of the dam proceeds, albeit amid widespread local and international protests.

Although the problems of the high dams are well known, other Himalayan countries continue to pursue them in order to meet their energy needs. In neighboring Nepal, which has over 6,000 rivers and streams, the hydropower potential is considered to be one of the highest in the world. With only about 10 percent of the population having access to electricity, Nepal considers energy development to be a main priority. The theoretical energy potential of Nepal, based on assessments of its rivers, is 83,000 megawatts, enough to meet the needs of over 700 million southern Asians. But the current installed capacity is only 227 megawatts. Clearly, the opportunities for hydroelectricity in Nepal are great.

Proposed large dam and run-of-the-river projects focus on the Mahakali and Karnali rivers in the west, the Kali Gandaki River in the central region, and the Kosi River in the eastern mountains. The proposed 270-meter-high dam at Chisopani, on the Karnali River, would have a capacity of almost 11,000 megawatts. That is enough energy to meet all of Nepal's domestic needs and allow some to be sold to India. The Kosi high dam, located in the eastern region, would generate an additional 3,000 megawatts, and the Mahakali project would have a capacity of 2,400 megawatts. It is clear that Nepal looks to electricity as one of its greatest possible export commodities.

But if all the proposed hydro projects in Nepal actually were to be built, over 2,000 square kilometers of land would be under water, including 20 percent of the country's total irrigable land in the hills.[16] Hundreds of communities in the mountains would be flooded, forcing the resettlement of hundreds of thousands of villagers. Nevertheless, Nepal intends to develop its hydroelectric potential for export sales, even if that causes major cultural and environmental disruptions.

The various problems that confront the big dams in Nepal are similar to those in India: the massive costs, borne by foreign assistance; the displacement of people from their villages; the loss of farmland and forest; and the safety risks due to seismic hazards. The Chisopani high dam on the Karnali River, for example, would displace about 60,000 people and submerge 28,000 hectares of farmland and forest.

A medium-sized hydropower plant along the Sanghla River provides small-scale electricity generation for local use. Photo by D. Zurick

All the Nepalese high dams will require massive levels of donor aid and international cooperation if they ever are to meet their potential. That may be especially difficult in light of the heavy-handed politics of Himalayan resource development, as the experiences of hydropower management in western Nepal have already shown.

In sharp contrast to the big dams, the micro hydropower schemes are owned by local communities and generate only about 200 kilowatts of electricity, enough to meet the simple electricity needs of a small village but not enough to export abroad. Many of the alternative development plans in the Himalaya include some aspects of small-technology energy development. By the early 1990s, over twenty government-run mini plants were operating in Nepal, with a total capacity of 950 kilowatts. Eleven more are on the drawing board. Similar micro projects have been initiated in northern Pakistan by the Agha Khan Rural Development Support Program; in northern India, especially in Ladakh (where they have the support of the Ladakh Ecology Project); and in Bhutan, where micro hydropower is being developed as a cornerstone of the country's rural energy program.

Mines and Industry

The resource wealth of the Himalaya extends to the very bedrock and the minerals it contains. Much of the work by the Geological Survey of India involves mineral exploration in Ladakh and Zanzkar, where rich sources of ore and precious metals are believed to exist. The military roads that were recently opened to the general public now put the mining areas within reach of industry. In the lower Siwalik range, near Dehra Dun and Mussoorie, the extensive mining of limestone already occurs, but with serious environmental impacts.

The commercial minerals in Nepal are thought to be mainly low-grade iron ore, scattered copper deposits, zinc, and limestone. It is possible that rich deposits of precious gemstones also exist in the Great Himalaya zone, but those have yet to be tapped. A new mine opened beneath the broad shoulder of Ganesh Himal, at over 5,000 meters elevation, is ostensibly for zinc production but precious metals and gemstones might also be obtained there. A new road to the mountain mine has been built above the village of Dhunche, in the upper Trishuli valley, prompting all sorts of rumors about sapphires, rubies, and riches.

Sikkim contains significant deposits of metallic ores and minerals, but they are not easily exploited, owing to the absence of roads into the mining areas. The Sikkim Mining Corporation, in cooperation with the government of India, oversees the mining activities at the new Rangpo Copper Mine, which has shown considerable potential for development. Additionally, Sikkim has important deposits of coal and graphite, both of which are now mined. Initial mineral explorations in Bhutan show the country to contain significant resources of dolomite, limestone, coal, and gypsum. Their extraction, however, remains minor; in 1990 total mineral production in Bhutan constituted less than 1 percent of the country's gross domestic product.

The mineral wealth of the Himalaya, although relatively undeveloped to date, constitutes a major aspect of the overall industrial development in the mountains. It joins with the hydropower generation and the extraction of primary resources such as forest products and agriculture to promote industrial growth in the region. Indeed, the mobilization of resources and labor for industry is a major force in the transformation of Himalayan livelihoods. Foreign trade, entrepre-

Mining industry, Sikkim. Photo by P. P. Karan

neurial activity, and government policies are additional instruments
in the expansion of industry in the mountains.

A concern for environmental quality cautions against unbridled
industrial growth, but it ranks low in the minds of most government
planners. New forms of pollution control have been put into effect in
some Himalayan towns, but places such as Kathmandu, which now is
heavily polluted from brick factories and vehicle exhaust, show that
the rules are rarely enforced. Various levels of environmental contamination afflict other industrial towns in Nepal, as well as in the Doon
Valley and the Indian hill towns of Shimla, Darjeeling, and Gangtok.
Air and water pollution were unheard of in the Himalaya only a few
decades ago, but they now have become widespread features in the
landscape.

In addition to the various kinds of horticulture promotions and
the new factories, most of the Himalayan development plans also include broad-based tourism goals. Tourism is championed by the economic planners, the conservationists, and the sustainable-development proponents because it is thought to be a light industry that can
perhaps balance the needs of conservation and economic growth. The
recent establishment of mountain parklands, as well as the extension

of conventional tourist facilities such as hotels, transportation, and restaurants portend the continued expansion of mountain tourism. Such initiatives have added to the building frenzy that goes on in many mountain localities, including now some of the most remote villages in the world.

This Darjeeling hill resort in the eastern Himalaya is a popular destination for Indian tourists (Mount Kanchenzonga is in the background). Photo by P. P. Karan

A Mountain Theme Park?

PEOPLE HAVE BEEN VISITING the Himalaya for a very long time. Some of the earliest travelers to the region were religious pilgrims on a *yatra*, or sacred journey. They sought among the holy mountains the personification of their Hindu or Buddhist deities. Their travels, deeply embedded in spiritual practices, inscribed large-scale religious circles, in which holy spots in the mountains were tied to southern Asia's giant mandala of sacred space. A network of pilgrimage routes connected the Himalayan spots with the religious geographies of India and Tibet. The spiritual landscape of the Himalaya was tied as well to the Hindu and Buddhist calendars, so travel occurred during auspicious times of ritual celebration. The Maha Shivaratri festival in Nepal, for example, brings tens of thousands of devotees to Pashupatinath Temple during the midwinter full moon to celebrate the night of the Himalayan deity known as Shiva.

14

Holy spots in Garhwal, especially the source of the Ganges River near Badrinath and such famous pilgrimage places as Gangotri, Kedarnath, Haridwar, and Rishikish, continue to attract millions of religious people every year to the western Himalaya. They arrive on foot, by bullock cart, in private sedans, and aboard video tour buses. Farther to the east, in Nepal, the holy springs and fires at Muktinath, located high in the mountains north of Annapurna, attract devout pilgrims who walk for weeks for ritual ablution. Other holy places of travel in Nepal that continue to attract spiritual travelers include Lumbini, the tarai village where the Buddha was born, and the Buddhist center of Bodhinath. Scattered elsewhere in the Himalaya, from the Indus to the Brahmaputra River, are countless other places of local worship. When travelers visit these spots, they enter Dev Bhumi, or the land of the gods.

With the rise of the British colony in India during the early nineteenth century, a seasonal exodus of Europeans to the mountains began. Colonial administrators, merchants, and military and their families all flocked to the high elevations, not for religious purposes but to escape the summer heat in the plains. They built summer residences in Shimla, Mussoorie, Nainital, and Darjeeling, and their cottages, riding stables, gardens, and pubs imitated those of England. In Kashmir, where the local maharaja banned foreign ownership of land, the British built luxurious houseboats and lived on the lakes. The hill stations eventually became important administrative centers as well as holiday resorts. Nowadays, they are the centers of mountain tourism.

Beginning about 1970, travelers from Europe and North America began to arrive in the Himalaya in steady numbers. They sought in the mountains the elusive ingredients of adventure, exotica, and a release from the conventions of Western society. The early travelers looking for adventure first came overland across Asia after an arduous journey, but they soon were replaced by legions of Western tourists, who arrive now by airplane in tour groups. The new tourists visit the Himalaya to trek, climb, river raft, or bicycle amid the highest and most spectacular mountains on the planet. Adventure travel packages are offered to tourists throughout the Western countries. They lead people into some of the world's most remote terrain. The mountains provide the visitors with the grand scenery of snowfields, rocky peaks, glacial valleys,

and panoramic views. But it is not a wilderness landscape. The people who inhabit the mountains are equally fascinating to the tourists, who tend to measure the exotic Himalayan cultures against their own ways of living. The intrigue of the natural scenery and living cultures entice so many visitors to the Himalaya that tourism has become one of the region's greatest industries.

Tourism is attractive mainly because of its economic potential, but it is more than simply a means of exchanging money. For the travelers themselves, the Himalaya offers a chance to literally visit the ends of the earth, a means to achieve a sense of far-off places, and a new way even of measuring life. According to the industry brochures, the adventure packages do not sell "tours," they create timeless travelers and enact nothing less than mythical journeys into places that test the very mettle of life. But in their mountain destinations, which initially provided the qualities of geographic remoteness, aesthetic landscapes, and authentic cultures, the tourists precipitate the very conditions of modernity that they seek to escape. Tourism forever changes the places where it occurs. Here it connects the remote frontiers of the world to the global economy and delimits both a new time and space. Tourism may offer new economic possibilities, but often at the loss of cultural identity. And in their search for the authentic, the tourists trample the very ground beneath them. The impacts of tourism on both society and the environment must be measured against its economic promise. Perhaps the most alarming paradox of Himalayan tourism is that in its quest for the ultimate adventure, it turns extraordinary mountain places into commonplaces.[1] Tourism is forever trying to escape from itself. The irony is that in a mountain land ruled by Hindu deities, where Shiva is king and the laws of karma still govern the world, the promise of tourism may in the end become a masterful illusion.

Tourist Tracks in the Mountains

There is no doubt that tourism brings money. It is for that reason that the Himalayan countries entertain major tourism initiatives in their national development plans. The International Center for Integrated Mountain Development established a major new focus on tourism in 1995. Its intent is to discern how tourism might be developed in the villages and how to ascertain the actual tourism carrying capacity of mountain places. The idea is that certain places can accommodate only a limited number of visitors before they are drastically changed.

Most tourists who visit the Himalaya come from the Indian plains. That fact is lost on many people who consider tourism to be a phenomenon mainly of Western society. Indian tourists include the pilgrims, who make up the bulk of the travelers; the visitors to the historical and cultural sites; and the people who wish to beat the heat of the summer in the plains by escaping to the mountains. In recent years, many Indians have come to the Himalaya to trek and to fish or to pursue other outdoor activities in the mountains. Despite their large numbers, little is actually known about the Indian tourists. The primary economic value of tourism lies in its potential to generate foreign earnings, so the affluent tourists from Europe and America tend to be counted the most.

The western Himalaya receives by far the largest numbers of tourists in the range. In 1993, Himachal Pradesh reported 3.5 million visitors. Shimla hosted more than a quarter of them. That hill town is especially popular during April to June, when middle-class Indian families flock to the mountains to escape the prickly heat of summer, just as did their colonial predecessors. The booming tourism business in Shimla has led to drastic new building and construction, so much so that geologists fear the town is sinking under its own weight. Other popular destinations in Himachal Pradesh include the hill stations in Kangra, the Tibetan community of Dharamsala, and the resort town of Manali in the Kulu Valley. Kashmir used to get about 100,000 visitors a year, but the violence there has directed tourists to Himachal Pradesh, instead, and to the Rohtang Pass, which crosses into Ladakh. In the neighboring Garhwal and Kumaun regions of Uttar Pradesh, a staggering 14 million tourists visit the mountains every year. About two-thirds of them are pilgrims. Less than 1 percent of the tourists in these regions are foreigners, compared to 6 percent in neighboring Himachal Pradesh. More than a third of the Indian tourists come from faraway Bengal or from southern India; the Himalaya has managed to maintain its all-India appeal through the ages.

Tourism in Nepal, on the other hand, is mainly a Western phenomenon. Hardly any Nepalese citizens are tourists in their own country. Indian tourists composed only a third of the 350,000 visitors to the kingdom in the mid-1990s. Most of the tourists came from Europe and North America or from elsewhere in Asia, such as Japan or Hong Kong. Nepal's primary attraction is its unique geography. The country straddles the world's highest mountains and the major cultures of India and Tibet, providing for the tourists the magnetic features of unsurpassed alpine scenery, quaint mountain villages, historical sites, and wildlife parks. If that is not enough, tourists can find in Kathmandu five-star hotels, gourmet cuisine, and gambling casinos. The latter especially attract Indian tourists, who make up 85 percent of the casino clientele.[2] Many of them arrive in Kathmandu on gambling package tours.

Nepal's tourism plans are ambitious. The country promoted 1998 as The Year of the Tourist. By the beginning of the new millennium, government officials want to see 1 million tourists in the kingdom every year. An enormous amount of the construction under way in Kathmandu is geared toward the hotel and restaurant industry. If the projections fail, there will be a lot of empty rooms in town. Villages situated along the main trekking routes have also invested heavily in tourism, anticipating ever-increasing numbers of foreigners walking in the mountains.

Not a lot of information is available about tourism in the far eastern regions. According to the Sikkim Tourism Office in Gangtok, the numbers of Westerners to visit Sikkim increases at a steady pace each

Tourists shop for handicrafts in Kathmandu. Photo by D. Zurick

year. The Department of Tourism now arranges package tours from Darjeeling or Siliguri, via the Sikkim Nationalized Transport Services. Buses take visitors into the mountain regions around Gangtok and to such prominent monasteries as Pemayangtse, Phodong, and Rumtek. Bhutan, meanwhile, pursues a policy of controlled tourism under the development of the quasi-independent Bhutan Tourism Corporation. The number of Western tourists in the kingdom was only a couple of thousand in the mid-1990s, having remained stable since 1987, when the country received a high of 2,524 tourists. The kingdom placed a cap on tourists in 1998 at 4,000 persons. In keeping the numbers low, Bhutan hopes to avoid some of the problems associated with full-blown tourism in Nepal. Finally, international tourism in Arunachal Pradesh is basically nonexistent. Except for a few areas in the valley of the lower Brahmaputra River, the region is off-limits to foreigners.

Tourism and the Economy

The potential contribution of tourism to economic development in the Himalayan countries is considered to be great and is the chief reason for its widespread promotion in the region. Tourism in Nepal provides about one-fifth the country's total foreign exchange. In Himachal Pradesh, tourism earnings are estimated to be about 2 billion rupees. Even Bhutan, despite its restrictive policy, reported that tourism earnings accounted for 5 percent of the gross domestic product in 1990.[3] The numbers of tourists may be low in Bhutan, but they spend quite a lot of money there. Tourism incomes are offset, however, by the large capital outlays required to support the development of the tourism sector. Hotels, transportation, and other investments are needed to manage the fledgling industry. Many imported items, such as food and furnishings, are needed to serve the tastes of tourists. The costs of such supports can be high, requiring significant national investments. The development plan in Himachal Pradesh for example, allocated over 18 billion rupees for tourism development purposes.

The fact that overall earnings from tourism may be less than what is anticipated by the planners is due in part to the costs of the tourism investments, but it also reflects the international character of the tourism industry, which is managed mainly by agencies located abroad. The way it works now, much of the tourism money never really enters the mountains. Most of the package tours for mountain trekking, for example, are bought and paid for in Europe or North America. The money tends to stay in the Western countries where the bookings are made. The amount that eventually reaches the Himalayan countries becomes concentrated in the capital cities or large towns, where the hotels and guide services are located. The villages, meanwhile, receive the bulk of the tourists but little of their money.

Because tourism has a fairly high start-up cost, in the villages wealthy persons tend to be the ones who benefit from it the most. The rewards for the others are minor. In the Kulu Valley, only 10 percent of the jobs created by tourism are filled by local people; other employees come from outside the region. In some parts of the Indian Himalaya, more than half of the total tourism business is controlled by outsiders. In the trekking business, mountain porters do the hardest work but get paid the least; in Nepal the daily porter's wage is $3. In Pakistan, the porters have organized to demand higher wages. As a result, though, the Pakistani porters in Hunza and Baltistan are labeled as agitators by the tourism industry and shunned by tourists, who see them as troublemakers.

A recent study on Himalayan tourism conducted by the International Center for Integrated Mountain Development in Kathmandu found that very little attention was given by tourism planners to strengthening the economic ties between villagers and tourism and that few incentives were provided to recycle tourism earnings back into the villages.[4] The main problem with the economics of tourism is that it produces greater inequities in the rural places. By linking villages with the international tourism economy, and thereby promoting a greater dependence on the world system, tourism also may, in the end, further erode the sustainability of village life.

Tourism and Mountain Society

As tourism forces contact between villagers and foreigners, it brings to the Himalaya the dilemmas of modern life. Under ideal circumstances, that meeting is managed to minimize disruptions in traditional life, which is, after all, what most tourists want to witness. Bhutan accomplishes this by strictly regulating where tourists may go, in what numbers, and how they may interact with local inhabitants. While some tourists may view this as being rather repressive, it is Bhutan's attempt to mitigate the cultural impacts of tourism. Bhutan, though, is an exception. More commonly, the Himalayan countries want as many visitors as possible, and their tourism policies do not really consider the full consequences of tourism growth on village society. In many cases, its social impacts are still unknown.

Some of the changes brought by tourism are superficial: new clothing styles and food preferences; cassette players and rock-'n'-roll music; burritos and hamburgers; and so forth. Such changes can probably be expected regardless of tourism. Other shifts, though, are more fundamental to mountain life. Because it tends to view culture as a sort of commodity or resource, tourism may actually cause a change in cultural identity, a gradual erosion of cultural values, new meanings of social interaction, and even the formation of new worldviews. Those are big changes. According to anthropologist David Greenwood, "Treating culture as a natural resource or a commodity over which tourists have rights is not simply perverse, it is a violation of the peoples' cultural rights . . . the commoditization of culture in effect robs people of the very meaning by which they organize their lives."[5]

Tibetan villagers rehearse a dance for a cultural ceremony at Tabo Monastery. Photo by D. Zurick

Not everyone shares Greenwood's perspective, though, and in the Himalaya many of the proponents of tourism think that its economic potential far outweighs the social disruptions. The Nepalese geographer and former Tourism Minister Harka Gurung believes that the primary problem in the Himalaya is poverty and that tourism should be aggressively pursued without sentimentality about the possible cultural consequences. His view is widely shared by planners throughout the region, as evidenced in the vigorous promotion of tourism across the range.

Some advocates of rural tourism argue that native cultures will creatively absorb tourist demands and that tourism can be a force to preserve culture in the midst of other modernizing trends. Tourists, after all, want to see things as they were, not as they may become. There is some evidence to support that viewpoint. For example, the heavy tourist traffic in the Khumbu region has trampled some of the mountain trails but not the Sherpa culture, which has proven resilient to many of the pressures of tourism. The Khumbu Sherpa are famous for the fact that they live near Mount Everest and produce world-class mountain guides. Most of the time, though, they graze sheep and yak and grow potatoes and buckwheat. They have managed to keep these livelihoods intact despite the overwhelming force of tourism in Khumbu.[6] One of the main aspects of Sherpa culture is reciprocity, the sharing of life. Some observers suggest that it has not been adversely affected by tourism. But other aspects of Sherpa culture have undoubtedly changed. Their worldview, for example, has shifted to accommodate the consumer model of the West. Their religious ceremonies have become secularized, shown for money to tourist groups. Their relations to the land and to the timing of work practices are transformed now, as such new duties as those performed by guides, cooks, innkeepers, and porters take on special meaning.[7]

The overall impact of tourism on Himalayan society may be impossible to fully assess, given the fact that its consequences may not be truly felt for generations. The potential of social impacts is known at a generalized level and even mentioned in most tourism policies, but little specific planning exists to deal with it. It remains an unknown quantity. The creation of new jobs predicts new social roles as well as new economic structures. Other lifestyle changes may also occur, and the overall shifts in values that accompany tourism remain largely unpredictable.

Tourism and the Environment

We have seen where the fragile character of the Himalaya may become stressed by the demands of population and industry. Tourism adds to those pressures by placing new requirements on land and water, by impinging upon established resource rights, and by producing conflicts in local environmental management. The environmental problems commonly linked to tourism include fuel-wood cutting for cooking meals and heating tourist lodges, timber cutting to build new tourist structures, trail-side litter and soil erosion, water pollution, and general habitat destruction. Land degradation is especially great around the hill resorts such as Shimla, Srinigar, and Mussoorie. These places have seen a recent boom in hotel construction and tourist traffic. It is evident also in the new tourism centers near the national parks and other scenic areas. Environmental damage has become common at the pilgrimage destinations in the high mountains of Garhwal. During the important religious holidays, huge crowds visit the temples there, causing damage to nearby vegetation and creating local sanitation problems.

In Nepal, serious environmental problems exist along the heavily used trekking routes. The Annapurna region in the central part of the kingdom, the Khumbu in the vicinity of Sagarmatha (Mount Everest), and the Langtang Valley north of Kathmandu receive about 90 percent of the trekkers in the country. All these areas report environmental damage due to tourism. Other congested places in Nepal, especially the urban centers of Kathmandu and Pokhara, also show the wear of tourism. The tourists in Pokhara are concentrated around the lakeside embankment of Phewa Tal. Their numbers have grown at a staggering rate, with little apparent control or regulation, and these have put tremendous pressure on the town's infrastructure and on the ecology of the lake. The pristine nature of Pokhara, which includes ready access to the nearby mountains, has been jeopardized by the presence of mass numbers of tourists.

In addition to the direct environmental consequences of tourism are many lesser-known impacts. The increased wealth that tourism brings to Sherpa people has allowed them to increase their numbers of livestock. The larger herds, in turn, cause overgrazing in some pastures. That is true also for areas in central Nepal and in Ladakh. The seasonal employment of tourism draws workers away from their farms at key times in the agricultural calendar, adversely affecting the maintenance of fields against such problems as soil erosion and pest infestations. Since it most often is men who leave for outside tourism work, women in the village work double time. The absence of workers also affects the maintenance of village trails and the exchange of labor that accompanies many traditional occupations.

Land-use changes may also result from tourism. Vegetables, fruit, and other luxury crops are grown for the tourism market, displacing in some places the farm production of grains and other food for local consumption. That is evident, for example, along the trekking routes in Nepal, where specialty crops and orchards supply the exotic items found on tourist menus in the lodges. The full consequences of such land-use changes are not yet known, but it is possible now to eat lasagna and apple pie in even the most remote villages in the Himalaya, and that in itself says a lot about the food that is grown and why.[8]

Many of the tourism policies of the Himalayan countries seek to somehow resolve the environmental and social impacts of tourism while still maintaining its economic potential. The design of ecotourism to replace mass tourism, for example, is thought to engender a more benign form of tourism growth, one that accommodates both environmental and cultural concerns. Throughout the mountains, the parklands and protected areas have been established as places where the diverse needs of economic growth, environmental conservation, and indigenous land rights may be reconciled. These places now contain extensive areas of land that is considered to be critical for the purposes of biological diversity, cultural survival, and tourism.

Parks and People

Almost 10 percent of the Himalaya is set aside for conservation under existing or proposed protected areas. Many of the big parks, such as the Great Himalaya National Park in Kulu, the Annapurna Conservation Area in central Nepal, and the Kanchenzonga National Park in Sikkim, are located in the high regions of the Great Himalaya. They are some of the most scenic and remote landscapes in the entire range. But very little habitat in the densely settled middle mountains is protected by parks. The proposed Black Mountains National Park in Bhutan will be a crucial addition to the preservation of temperate forests in the threatened middle-mountains region. Khaptad National Park in western Nepal also sets aside middle-elevation forest and pasture. Otherwise, most of the middle-mountains range is excluded from the Himalayan park system. Numerous refuges, on the other hand, are found in the low elevations of the outer foothills and piedmont plain. They provide important habitat for endangered plants and animals and are very popular with the tourists.

Of the protected areas in the Himalaya, most fall under the jurisdiction of a single country. Three large parks, though, cross international boundaries and form a new, transnational park type: Manas Park, astride the Bhutan-India border; Khunjerab Park, joining Pakistan and China; and Sagarmath-Qomolongma Park, circumscribing Mount Everest in Nepal and Tibet. Their sponsors all recognize that in order to achieve biodiversity and conservation management goals, the park boundaries should conform to ecological rather than strictly political criteria. That viewpoint, though, is difficult to get across, especially among entrenched politicians, so the transboundary parks are the exception rather than the rule. Most parks in the Himalaya, instead, reflect national economic priorities rather than international conservation goals. Therefore, the west to east distribution of Himalayan parks and protected areas, as recognized by the International Union for the Conservation of Nature (IUCN), follows along the lines of country interests. In addition to Khunjerab National Park, the Indus basin in Pakistan contains numerous small wildlife sanctuaries and game reservoirs. Most of them are located in the Karakoram range, though, and lie outside the Himalaya proper.

East and south of the Indus River in Kashmir are situated three national parks. The largest of them is Hemis National Park, located in the trans-Himalayan district of Ladakh. It covers an area of 4,000 square kilometers of high desert country. The other two national parks contain a combined area of 500 square kilometers. In addition to the national parks, over forty sanctuaries and game reserves of varying sizes occupy another 8,000 square kilometers in Jammu and Kashmir. Most of the game reserves have no assigned IUCN status and exist as parks on paper only.

Himachal Pradesh contains two large national parks—the Great Himalaya National Park, above the Beas River in Kulu, and the Pin Valley National Park in Spiti. The two parks join via the Rupi Bhabha Sanctuary to form a contiguous territory of 1,413 square kilometers. This vast highland park encompasses an extraordinary alpine region of temperate forest and cold desert. It contains diverse endangered

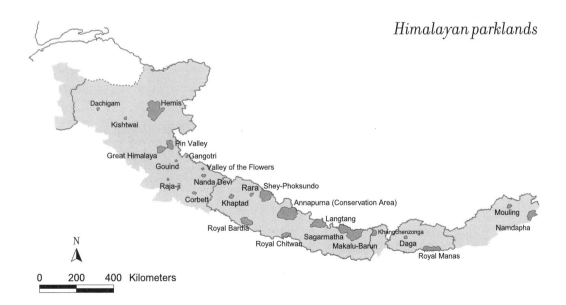

Himalayan parklands

wildlife, including the Himalayan brown and black bears, the musk deer, and the snow leopard. The integrity of the Rupi Bhabha Sanctuary is threatened, though, by the Sanjay Vidyut Hydroelectric project, which is a part of the Sutlej developments; it encroaches onto the sanctuary land with its maze of tunnels, roads, dams, and reservoirs.

In addition to its national parks, Himachal Pradesh contains twenty-nine smaller sanctuaries, which altogether preserve some 3,882 square kilometers. They include the 1,000-hectare Shimla Water Catchment Sanctuary, which protects the water supply of the town of Shimla. Originally the protected property of a local rana, the Shimla Sanctuary was established in 1958 as a preserve and may well represent the last undisturbed middle mountain forest in the entire front range of the western Himalaya. The Shimla Ecology Project now seeks to promote the ecotourism potential of the sanctuary by establishing walking tours and interpretive programs there.

In Garhwal and Kumaun, seven national parks totaling some 4,000 square kilometers and five sanctuaries with a combined area of 1,809 square kilometers dot the landscape. An additional national park at Kedarnath as well as nine new sanctuaries are also proposed for the region, for an estimated grand total of 10,000 square kilometers of preservation lands. The Garhwal parks include the magnificent 63,000-hectare Nanda Devi Park, which was designated as a World Heritage Site in 1988. The area is a vast glacial basin divided by north to south ridges and dominated by the 7,434-meter Nanda Devi peak. The designation of Nanda Devi as a World Heritage Site by the United Nations extends the ancient veneration of the mountain by Hindus, who see it as the embodiment of Shiva's consort Parvati, to the modern age of global environmental appraisals.

The kingdom of Nepal contains twenty-two existing or proposed protected areas covering about 17,000 square kilometers. Nine of those places are national parks, two of which—Royal Chitwan National Park and Sagarmatha National Park, are also World Heritage Sites. The established national parks in Nepal altogether cover 11,000 square kilometers. The Annapurna Conservation Area, located in the central part of the country, and the Makalu-Barun Conservation Area, adjoining Mount Everest, contribute an additional 6,330 square kilometers to Nepal's park list. Most of the protected areas in the country contain diverse human settlements whose populations must somehow be included in the design and management of the parks.

The important lowland wildlife reserves, such as Koshi Tappu, Parsa, and Royal Sukla Phanta, are all located in the Nepalese tarai. They protect the forest and riverine habitat for endangered wildlife species, including the Indian rhinoceros, the royal Bengal tiger, and the Ganges River dolphin. The tarai reserves also serve as crucial nesting areas for wintering birds from Tibet and as a resting site for avian migration across much of central and southern Asia.

Along its eastern border with Nepal, Sikkim established the 850-square-kilometer Kanchenzonga National Park. It was created in 1977 to protect the undisturbed forest and spectacular wilderness around the third-highest mountain in the world. However, the Yuksum area beneath Kanchenzonga, considered to be sacred land by the native residents, is marked for hydroelectric development. That enterprise has aroused strong opposition from the local Lepcha tribes. Sikkim contains four other sanctuaries, which in total preserve another 125 square kilometers. Most of those areas are less than ten years old.

Sikkim's neighbor Bhutan has set aside 20 percent of its land area for protection from development in a system of fourteen parks, reserves, and sanctuaries. They range in size from the five-hectare forest reserve at Zhoshing near the Sikkim border to the 658-square-kilometer Royal Manas National Park, located in the central Duars. Most of Bhutan's natural-reserve areas exist mainly on paper, though, without proper staffing, background resource surveys, or field management. Their legal status does prohibit hunting, poaching, unlicensed fishing, and other destructive practices, but the enforcement is not yet very strong. Despite these limitations, the designation of so much protected land in Bhutan suggests a fairly enlightened approach to combining conservation needs and economic growth.

Finally, in the far eastern sector of the Himalaya, Arunachal Pradesh contains two national parks and four sanctuaries. Altogether, they preserve 4,000 square kilometers of mountainside. An additional nineteen national parks and sanctuaries are proposed in the state. Most of those areas are located east or south of the Brahmaputra River and so fall outside the mountain range. However, the Mouling National Park in the Mishi Hills and the Mehao Sanctuary near Pasighat contribute 750 square kilometers of protected land to the

Himalaya.[9] The far eastern Himalaya is the last true outpost of Asia, and its preservation is key to safeguarding some of the world's most important biotic resources.

Protected Areas and Ecotourists

The objectives for the Himalayan parklands are diverse, as shown by various management styles and by the complicated regulations that govern them. In most cases, the parks are central elements in the sustainable development of the mountains. That fact highlights the value of the parks for both biodiversity goals and resource management. The latter includes diverse economic programs, including tourism. Furthermore, the park governments attempt to accommodate the land rights of native people who live within their boundaries.

The setting aside of land for conservation is a relatively new approach to the environmental problems of the Himalaya. The formation of protected areas is based on the concept of multiple land use, in which the needs of environmental protection are balanced against those of local people and national development. Although theoretically satisfying, this complicated strategy is difficult to manage on the ground. In many cases, the arguments in favor of parks emphasize their economic potential. That is understandable given the fact that the top priority of the Himalayan countries is to decrease poverty. There is risk, though, to such an evaluation; the conservation objectives of the protected areas may become secondary to those of tourism or other economic activities.

In most cases, the appropriate type of tourism for the Himalayan parklands is thought to be ecotourism. In the ecotourism model, adventure places are created which attract tourists because of their natural or cultural qualities.[10] Such places include remote highlands and relatively undisturbed lowland jungles. These are vastly different worlds, but they both exhibit interest-provoking features of scenery, wildlife, and local culture. Some of the parks remain undeveloped; for example, the Pin Valley National Park, known best for its difficult access, the presence of snow leopards, and the absence of tourist services. Others have become full-blown tourist resorts. Tiger Tops Resort in Nepal, located in the middle of Chitwan National Park, boasts luxury accommodations, swimming pools, and all sorts of conveyances into the surrounding jungle for wildlife viewing.

The nature-based ecotourism programs include various adventure opportunities such as river rafting, elephant rides, trekking and climbing tours, mountain biking, and even hot-air balloons. These activities now take tourists into the most remote park territories. The adventure excursions are questionable according to conservation criteria, but they do bring in some money. Hence, they are promoted all across the Himalaya, as well as in the back pages of the environmental and tourism magazines published in North America and Europe.

Two examples from Nepal, the Annapurna Conservation Area Project (known as ACAP) and the Sagarmatha National Park, illustrate the prospects and problems of the Himalayan protected areas. ACAP was established in 1986 under the auspices of the King Mahendra Trust for Nature Conservation, the kingdom's premier environment organization. At the time of its official gazetting, ACAP was to manage comprehensively 4,000 square kilometers around the mountains of Annapurna. In addition to the magnificent natural scenery of the area, the ACAP project area contains 40,000 people living in over 300 farming villages.

A number of environmental regulations were put into place in the region after consulting with local villagers. Restrictions were placed on livestock grazing and forest cutting, pollution controls were formalized, and education programs were conducted in the villages on such topics as environmental conservation, energy, and tourism. Because ACAP from its conception involved the local people in its design and management, it has managed to avoid some of the conflicts that plague other Himalayan parklands. Notably, all decisions regarding indigenous land rights and resource management are made by village participants working with park personnel. That is vitally important to the villagers, whose lives depend upon access to forests and pastures.

But ACAP intends some day to become fully self-supporting. Right now it exists mainly on donor assistance and government subsidies, as well as on visitor permits. In order to meet its financial needs, ACAP will need to expand its tourism base. The adverse consequences of that for the fragile uplands may well compromise some of ACAP's initial conservation goals. Field observation has shown that land degradation

Villagers meet to discuss issues of resource management in the Annapurna region of Nepal. Photo by P. P. Karan

is already apparent around the most heavily used areas, particularly along the trails near Gandruk, the project headquarters. Despite the difficulties in balancing conservation and economic growth, ACAP has been successful in managing the cooperation of native people and tourists and serves as a model for similar ventures elsewhere in the Himalaya.

The Sagarmatha National Park, on the other hand, was established without much consideration for the local Sherpa population and consequently has had to deal with ongoing conflicts between park managers, villagers, and tourists. The park was set up in 1976 in a 1,148-square-kilometer area just south of Mount Everest. It is a dramatic landscape of high mountains, deep valleys, and glaciers. Over 3,000 Sherpas live in villages in the park. Around the villages are small farm plots, and the forests contain pastures where livestock graze. Nowadays, most Sherpa also work in the tourism sector. Over 8,000 tourists visit the park each year, bringing with them about $75,000 U.S. in park admittance fees. That is only a portion of the tourism earnings, though, for it doesn't count the payments to guides, porters, cooks, innkeepers, merchants, and other entrepreneurs.

Sagarmatha National Park was established in Khumbu with the assistance of the New Zealand Park Service. It follows the Western park model, which is based on managing wilderness with no significant human settlements. There was little that could be learned from the Western models about how parks should accommodate the needs of resident people. Instead of integrating local villagers into the park, as ACAP did, they were isolated in small enclaves and kept out of decisions regarding park regulations.[11] That quickly led to poor relations between the Sherpas and the park personnel. Park managers were stymied by the lack of cooperation about matters such as forest use and pasture rights. The situation grew combative when the Sherpas discovered that their world in fact had been taken from their control. Some new programs were established in Sagarmatha in the 1990s which sought to bring villagers more directly into the sphere of park decisions. The park government has been somewhat successful in mollifying native concerns, but the Sherpas have not completely acquiesced to the new management styles and the situation remains somewhat tense between villagers and the park administration.

Annapurna and Khumbu are examples of how regions can manage the Himalayan worlds as parklands. They commonly inform other attempts to develop protected areas in the mountains. Makalu Barun, which adjoins Sagarmatha, Kanchenzonga in Sikkim, and the Great Himalaya National Park in Kulu base their designs to some extent on the models from Nepal. The more innovative parks recognize that in order to be effective in promoting sustainable development, they need to initiate conservation programs that have the support of local people and that will in some way provide economic benefits to them. Education and employment opportunities are necessary ingredients of park programs, as are conservation practices based upon scientific understanding. The conflicts between development and preservation, industry and agriculture, modernity and tradition will be around for quite some time in the Himalaya. The parks are thought to be one strategy that will work to resolve them.

PART FIVE

HIMALAYA IN THE TWENTY-FIRST CENTURY

A mountain landscape in the western Himalaya. Photo by D. Zurick

Landscapes of the Future

THE EXTRAORDINARY PROFILE of the Himalaya, seen from the distance of the southern plains, appears to be eternal. The snow-clad peaks stand firm against the deep blue of mountain sky, in dramatic testimony to the relentless power of drifting continents. They etch a stark silhouette on the horizon that has an almost crystalline quality. Whereas the summits rise stalwart and icy white above the middle mountains, the lower slopes are made green and softened by vegetation; from a great distance it is impossible to say if the color is due to the forests, the farm fields, or the pastures that cover the hillsides. The rivers that run through the mountains carve them into fantastic shapes before spilling water and sediments onto the plains.

From that distant perspective, caught in a single glance, the dynamic character of the Himalayan landscape is barely discernible: the

15

geological upheaval of the mountains and the weathering of the land are not apparent; the shifts in vegetation are blurred; the rivers seem to run evenly through the undulating terrain. The distant viewer might easily assume a natural stasis in the mountains, where the shape of the land would appear timeless. That view denies, of course, the compelling qualities of diversity and transformation that more aptly describe the Himalayan world. The mountains would seem also to tower above the pursuits of humankind, impervious to them. This latter image may be contained in the mythologies of resident cultures or in the promotions of the visitor industry, but it is contrary to the experiences of everyday life in the mountains, where humankind continually shapes the Himalaya to its needs and to the findings of modern scientific research, which show both alarming levels of land degradation and more hopeful signs of ecological renewal.

Natural forces are so dominant in the Himalaya that the mountains are described in geological terms as a high-energy environment, the bedrock of which shifts with the movement of the earth's continents. Resulting earth tremors may have devastating impacts on the mountain settlements. In the face of torrential monsoon rainfall, the steep slopes are easily eroded by soil creep and land slips. In severe cases, entire hillsides may disappear beneath mudslides that scar the countryside and bury villages. Highland lakes dammed naturally by glacial moraines may suddenly burst and flood the downstream valleys. The fast-flowing rivers that descend through the steep relief of the mountains are powerful agents of erosion. Such constant exposure of the land to the forces of nature shapes it into continually new configurations. All this activity diminishes the notion that the physical condition of the Himalaya is steadfast or somehow resolute, and it makes the mountains a risky place in which to live.

Along with the natural forces are the modifications of the land made by the people who live there. The pursuit of livelihood, deeply embedded in the needs and aspirations of human cultures, reshapes the Himalaya according to the designs of society. They, too, frequently change. Natural and social forces may not be easily distinguished from each other, and, furthermore, they are often unpredictable. It is understandable, therefore, that a clear and precise idea of environmental conditions in the mountains has remained elusive. "Fragility" is the word most commonly employed to describe the mountain environment, but more properly we could use the term "dynamic."[1] It implies a process rather than a condition.

We have so far seen that the environmental history of the Himalaya includes its own seismic past as well as the successive impacts of human society. The advance of modernity in the mountains gradually issues from the early historical periods of wandering tribes and Rajput princes, and it intensifies at a quickening pace with the subsequent eras of imperial control, colonization, and postcolonial development. The contemporary status of the mountain environments must be measured against these past events. They point to the pressures of growing human populations on the land, the concentration of power among outside interests, the reduction of local authority over resources, and the aggrandizement of commercial interests in mountain resources. The current levels of land degradation in the Himalaya are linked to all these things.

In the midst of such sweeping changes are situated the struggles of native people to retain their rights to territory and to the traditional use of local resources. The claims of mountain communities and the preservation of the natural environment have become crucial elements in the formulation of new sustainable-development outlooks, which seek to integrate the birthright of indigenous peoples with conservation programs that may allow economic development in the Himalaya to proceed in a responsible way. That tripartite goal largely defines the critical issues that confront the region as it moves into the twenty-first century.

The Affairs of States

Population density in much of the Himalaya has more than doubled since the 1950s. That has put a great pressure on forests and has contributed up to 50 percent of the decrease in per capita farmland across the mountains. Despite a reported slowing of population growth in some places, the generally high fertility rates in the Himalaya indicate that mountain populations will continue to grow at alarming rates. If the 1991 population growth rate is maintained, for example, the number of persons living in the Himalaya will double in less than thirty years. Under these circumstances, the scale of the mountain economies and the physical carrying capacities of the natural environment may well be surpassed.

Many of the conditions that already are considered to be critical in the region—for example, the stagnant nature of mountain farming systems and the growing intensity of resource use for subsistence purposes—will likely worsen as the population continues to grow. Hence, in discussions about the future of the Himalayan environment, the magnitude of population pressures on the land cannot be ignored. It is at the forefront of many national and regional planning proposals. An ironic and troubling aspect of the population problem is that the large family sizes that are prevalent in the mountains actually represent effective household strategies for dealing with the problems of dwindling land and resource supplies. The children in the villages provide a reliable pool of labor and increase the economic opportunities of the family; hence, many children are desired to avert household risk in uncertain times. Under such demographic imperatives, slowing down the rate of population growth will require careful attention to the social and ecological fabric of villagers' lives. Because of its social context, the problem of population pressure will not be easily resolved. Nevertheless, family planning programs are vigorously pursued in the Himalaya, in combination with efforts to increase productivity on subsistence farms.

Contemporary industrial activities in the Himalaya intensify resource use and extend the long history of resource appropriation in the mountains. The commercial felling of timber, the diversion of water for hydropower generation, the construction of roads and other infrastructures for the purposes of regional integration and market development, and the commodification of mountain agriculture all seek to enhance the economic productivity of the mountains which is crucial to the development of the national economies. Such activities, however, may have adverse environmental impacts that greatly outweigh in both scale and magnitude those of subsistence livelihoods. Moreover, industrial growth fundamentally changes the mountain societies; it promotes an orientation toward material consumption.

The pressures emanating from the subsistence economy and the commercial demands of modern development, along with the geological instability of the mountains, combine to produce the suite of environmental problems in the region. Much scientific research seeks to distinguish the relative contributions of human activity versus natural events to the problem of land degradation in the mountains, so that appropriate policies may be devised. The mountain countries, meanwhile, are concerned with how to improve economic standards without undue loss of their natural and cultural heritages.

BHUTAN

The emphasis of conservation development in Bhutan is on maintaining the current favorable status of that country's natural habitat. Unlike that of the other Himalayan countries, much of Bhutan's natural landscape remains largely intact; and the country seeks to establish sustainable environment management and economic development that will preserve its natural resources. But Bhutan faces several alarming environmental trends that may threaten the future of its natural heritage.

Regional demands for timber, growing at a rapid pace, will result in greater commercial felling of trees in Bhutan. This already is a problem where new roads make the forests accessible to timber cutters. The steady increase in the numbers of people and livestock will place additional stress on the country's forest resources. Where the two demands coexist, the conservation of natural areas is undermined. For example, much of the animal grazing in the forests already occurs along the roadways and paths; hence, where overgrazing is a serious threat highways and logging will further encroach upon the forests.

The relative scarcity of arable land in Bhutan predicts greater farming intensity regardless of food imports. That jeopardizes the stability of the steep slope lands if careful soil management is not practiced. The emergence of an industrial sector—for example, among the towns of the Duars Plain—and of that of growing consumerism in Bhutan are also potentially damaging forces in the country. The pressures they place on the country's limited resources and fragile ecology will increase proportionate to their role in the future Bhutanese economy.

In light of these trends, the environmental policies of Bhutan are straightforward and direct and provide a dose of optimism amid the now widespread environmental degradation in the Himalaya. In order to preserve the country's intact natural areas, a National Forest Policy was initiated in 1985 with the primary goal of keeping at least 60 percent of the country under forest cover. Economic objectives are secondary to those of conserving the natural resources in the forests. The government of Bhutan nationalized most commercial logging in the country in 1979, and while the rate of timber cutting has since increased, it follows a go-slow policy of selective cutting and reforestation. Comprehensive forest management plans now cover over 5,000 square kilometers, and in recent years tree planting has become a top priority.[2]

Despite its large potential, the development of hydropower in Bhutan has followed a modest path. The country's largest hydropower scheme is the Chukka project, which was completed in 1988 to generate 300 megawatts of power. It supplies the towns of Thimphu and Phuntsholing and a number of industries located in western Bhutan. The remainder of the energy is exported to West Bengal. The relatively small size of the project, measured against the Himalayan high dams and the fact that it contains no large reservoir, has minimized the environmental impact of the Chukka project. The rest of Bhutan's hydropower schemes are micro projects with a combined production of less than twenty megawatts of power. Firewood remains the primary energy source in the country, contributing over three-fourths of Bhutan's total consumption, but the country is developing alternative, renewable energy sources such as biogas, solar, and wind power.

The comprehensive goals of Bhutan's conservation approach to development are contained in the Paro Resolution on Environment and Sustainable Development put forth on May 5, 1990. According to the Paro Resolution,

> The Kingdom of Bhutan now stands at an important crossroads
> . . . We believe that preserving, indeed strengthening, Bhutan's
> natural resource base is central to a sustainable and prosperous
> future for the country. We urge the development of a National Environmental Strategy that will ensure the careful stewardship and
> sustained use of these natural resources . . . The key is to find a
> development path that will allow the country to meet the pressing
> needs of the people, particularly in terms of food, health care and
> education, without undermining the resource base of the economy.[3]

A lumber camp in Bhutan.
Photo by P. P. Karan

In keeping with the Paro Resolution, the emphasis of development in Bhutan is given to population control through comprehensive family planning and health services, to watershed management through balanced land-use planning, to preserving Bhutan's rich biodiversity through implementing restrictive forest policies, and to maintaining the cultural heritage of the country. The latter is especially important insofar as it is the basis for the Buddhist ethic that guides the country's philosophy of sustainable development. The formal environmental concerns are facilitated by various government sectors, organized under the Bhutan Trust Fund for Environment Conservation, which was created in 1991 to guarantee a sustainable and well-accounted revenue source for meeting the environmental goals of the country.

Land degradation in Nepal was reported to be so severe and wide-spread that the country became the archetype for the Himalayan eco-crisis model. As a result, Nepal has become something of a cause célèbre with international development agencies. Foreign aid directed toward the environment now inundates the country. Meanwhile, demographic pressures, fueled by the current high annual population growth rate, which is expected to exceed 2.5 percent by the year 2000, continue to tax much of the countryside.

The high densities of human and livestock populations put great demands on farms, forests, and pasture lands in many localities. Where they are most severe, they have resulted in reduced resource quality and increased human poverty. Already more than half the population living in the mountains is below the poverty line. In many cases, the response to such conditions has been the march of migrants out of the hills and onto the tarai plains, an outmigration that is expected to extend into the next millennium. Food production, meanwhile, continues to fall behind population growth, fodder deficits are expected to triple, and fuel-wood shortages will increase fifteen times by the year A.D. 2011.[4] Although these projections are necessarily tentative, they inform various government policies regarding the environment and development in Nepal.

The reported deforestation countrywide does not accurately reflect regional forest trends in the country. As we have seen, the pattern of land-cover change in Nepal is diverse, with some areas showing an expansion of forest cover in recent decades and others a loss. These forestry advances in the mountains are tied to government and village tree plantings, to rehabilitating degraded forests by restricting the use of them, and to household tree planting on private lands.[5] Tarai studies report that the highest rates of forest loss are due to land clearing by immigrants and forest clearing by the timber industry.

These mixed trends contain some hopeful signs of environmental renewal in Nepal. The more positive conditions generally occur where cultural traditions remain strong or where government conservation programs focus on people rather than simply on trees. Nonetheless, the overall land degradation in the country continues to pose serious threats to the efficacy of organic farming. The shortages of fodder and

Jute-mill workers in the Nepal tarai. Photo by P. P. Karan

the constraints on nutrient recycling are offset in some places by an increased usage of chemical fertilizers. In some limited instances, the use of fertilizer has reduced land degradation by lessening the demand for forest grazing. But fertilizer use is restricted to the most accessible places and to the wealthier farmers, and the chemicals have their own adverse impacts on soil and water quality.

In addition to the resource problems commonly reported for the subsistence economy, new environmental dilemmas in Nepal derive from contemporary urbanization and industrial development. Although Nepal remains one of the least urbanized countries in the world, the rate of urban growth has increased sharply in recent decades, from an urban population of less than 500,000 in 1971 to one of more than 1.5 million in 1991. The projected size of the urban population in the year A.D. 2000 is 4 million persons.[6] But the industrial sector of Nepal's economy remains relatively minor, accounting for less than 7 percent of the gross domestic product in the early 1990s, so jobs in the towns are hard to come by. The considerable constraints on industrial output include a lack of energy, raw materials, investment capital, skilled labor, and rural infrastructure. Without these ingredients, economic expansion outside of agriculture and tourism is limited. Still, the growth of the industrial sector is an important aspect of Nepal's development planning. A great deal of emphasis is now placed

Polluted water and garbage degrade the quality of the Bagmati River in the Kathmandu Valley, Nepal. The river is used for irrigation, laundering clothes, and bathing by many city dwellers. Photo by D. Zurick

on increasing foreign investments and on liberalizing the private sector policies. These efforts are intended to promote new factories. Without effective regulations, they will likely bring on more environmental problems.

Water and air contamination already exist in the tarai towns, in the Pokhara and Kathmandu valleys, and along the country's growth corridors. The Birganj-Hetauda route, which links the Indian border with the Kathmandu Valley, contains over three-quarters of Nepal's manufacturing plants. It has become a major factory corridor in a blighted landscape. Much of the remainder of the industrial expansion is concentrated in the Kathmandu Valley, where the population has trebled between 1971 and 1991. The medieval character of Kathmandu is quickly changing as it industrializes. Air and water pollution from industry effluents, brick factories, and vehicle exhaust hang over the city. The rivers that run through the valley, and which serve as sources of drinking water and irrigation, are deluged now with pesticides from agriculture, toxic chemicals from the carpet industries, untreated discharges from industries, and much of the seventy-five tons of solid waste generated each day by the city's residents. The cobblestone streets are literally awash in garbage before the cows and street sweepers help to clean them up each morning.

In response to the adverse environmental trends confronting the country's future, Nepal developed in 1993 the comprehensive Environmental Policy and Action Plan. It has five broad and largely interwoven objectives: to manage the sustainable use of natural resources, to balance development efforts and environmental conservation, to safeguard national heritage, to mitigate the environmental impacts of development projects, and to develop institutions that will integrate the environment and development goals.[7] It is worth noting that these objectives are meant not to separate environment and development issues, but rather to link them together within a coherent national strategy. The success of such a program depends, of course, on political will and on how carefully the specific projects become implemented.

The greatest prospects for successful resource management in the Nepalese countryside lie in granting the rural communities more authority over local environmental management. That approach runs counter, though, to the centralized power and resource appropriations that characterize most current government policies. Hence, new ideas and institutions are required to implement this marked change. The current forestry plans, which include community programs, take a step in that direction by linking forests, farms, and pastures in a comprehensive management plan. The specific environmental initiatives that result tend to be more holistic in conception and serve to further enhance the participation of villagers in public-land management.

In some instances, where the natural or cultural heritage is at particular risk—for example, in the designation of protected areas or the management of historical sites—the direct involvement of government agencies is needed. Such interventions extend as well to the management of the environmental impacts of urban and industrial development. Pollution abatement is possible only through regulatory enforcement, and infrastructure development, especially of roads, needs to guard against soil erosion. Since unchecked population growth will likely exacerbate the rural problems, Nepal's environmental policy includes various family planning and health protocols. The intention of these programs is to improve the health status of the current population while decreasing the rate of population growth; the two goals are considered to be inseparable. Finally, the Nepal Environmental Policy and Action Plan seeks to build appropriate institutions that will support and monitor the implementation of future development strategies.

INDIA

India controls much of the area between the Indus and Brahmaputra rivers. Its mountainous areas include Jammu and Kashmir, Himachal Pradesh, and the districts of Garhwal and Kumaun in the western and central portions of the range; Darjeeling and Sikkim located east of Nepal; and Arunachal Pradesh in the far eastern region. The diversity of environments and people across such a large span of territory is obviously great. The single largest problem in the region, from the government's point of view, is that of human poverty, which is driven by the lack of basic resources and the degradation of the land.

Official government reports still tend to identify land degradation in the Indian Himalaya with the impacts of the subsistence sector. Hence, the most commonly cited problems in the mountains are the conventional ones of forest exploitation for fuel wood and fodder, livestock pressure on pastures, and intensive farming on steep slopes. But the scope of industrial and commercial development in the Indian mountains clearly suggests that it, too, bears much responsibility for adverse environmental trends.

The development of roads, mines, hydropower, and cash farming in the Indian Himalaya produces significant foreign earnings for the region, so the environmental impacts of such projects are downplayed in the government reports. The officials in Himachal Pradesh regularly petition the national government not to let ecological considerations hamper economic development in the state. But India has recently made it mandatory for all agencies to complete environmental-impact assessments before carrying out development projects in the mountains, particularly those that affect water resources. Of course, the recommendations need to be followed or else they are merely additions to the stream of paper work that already clogs the government bureaucracies.

Recognizing the need for more information about the changing circumstances in its Himalayan territories, India has established several research and policy institutions in the region. Foremost among these is the G. B. Pant Institute of Himalayan Environment and Development, established in 1988 to acquire knowledge about environmental conditions in its mountain regions and how to improve the living standards there. The institute's policy areas include land- and water-resource management, sustainable development of rural ecosystems, environmental impact analysis, and conservation of biological diversity.[8]

Other collaborative Himalayan institutions in India include the Wadia Institute of Himalayan Geology, the Wildlife Institute, and the Indian Council of Forestry Research and Education, all located in Dehra Dun; the Council for Scientific and Industrial Research (CSIR) Laboratory, at Palampur in Himachal Pradesh; and the CSIR Institute at Jorhat in Assam; as well as over fifteen universities and research centers scattered across the Himalaya. The main concerns of these institutions comprise water management and soil conservation, sustainable farming systems, livestock management, horticulture development, tourism, and biodiversity conservation programs. The latter cover a number of the biosphere reserves, including the Nanda Devi and Manas national parks.

Much of the environmental policy in India remains under the control of state governments, which are responsible for the forests, agriculture, and energy development. Their agencies, in turn, are entrusted with the tasks of developing and implementing local environmental policies and with enforcing the regulations. To a large extent, these mandates follow along fairly conventional sector lines. But when economic returns are threatened by conservation rules, the former almost always prevail. In some instances, the state agencies have initiated quite innovative environmental programs. For example, the Department of Forest, Farming, and Conservation in Himachal Pradesh established in 1993 the Shimla Eco-Development and Conservation Project, which seeks to protect forest groves by permanent closures, to establish nurseries for seedling distribution to area farmers, to improve garbage disposal in the Shimla neighborhoods, and to establish a biodiversity center near Shimla that will function as a city-based ecotourism center. A strong component of the Shimla Project is community education and public outreach. The project seeks to enlist the participation of local villagers and townsmen in the establishment of its various programs. The practical outcomes of the Shimla Project, though, as with other similar programs located in the hill towns of Mussoorie and Darjeeling, are not yet known.

PAKISTAN

The mountain districts of Pakistan that fall within the bend of the Indus River are some of the most densely populated regions in the northern highlands. Exploitation of the natural-resource base there has increased dramatically in recent decades, leading to serious forest depletion and to deterioration of range lands in places such as Kohestan. The problem of population pressure is especially great in spots where refugees are relocated from Afghanistan. Pakistan still harbors over 3 million displaced Afghans. Moreover, the road extensions that have taken place in the mountain region upon completion of the Karakoram Highway make the forests more accessible for timber cutters. There has been a consequent increase in both licensed and illegal tree felling in the northern districts. Much of the Indus mountain territory is still controlled by tribal chieftains, limiting the enforcement of government rules and programs. The Agha Khan Rural Support Program, a nongovernmental agency with a strong conservation focus, seeks to promote sustainable development in the northern Pakistan mountains. It has had some limited success in working with tribal arrangements.

At a national level, environmental policies in Pakistan center mainly around resource development for economic advancement. To ease the pressures of local demand on resources, the government encourages the development of alternative economic opportunities in the mountains. These include horticulture and tourism, as well as small-scale industry. By its own admission, Pakistan tends to concentrate development activities among the irrigated plains or around the urban centers. Consequently, the level of investment in the mountains, apart from direct food relief, remains quite low.[9] In recent years, though, the country has established several policy initiatives focused specifically on the problems in the northern mountains. The 1992 National Conservation Strategy recommends an integrated development approach in the highland areas. The reforestation programs of the

1992 Forestry Sector Master Plan are targeted at the most severely degraded mountain districts. The Pakistan Agricultural Research Council established the Mountain Research Program to extend national food concerns to the highland farming systems. Much of its focus is on developing valuable mountain crops, such as temperate-clime fruits, for export and on improving the quality of grazing lands and pastures.

A Mountain Agenda

Environmental circumstances in the Himalayan countries combine to produce a regional dimension of land degradation that only recently has come to be understood. It is feared that the mountains, which have a worldwide appeal, are being ruined. The resolution of environmental problems is hampered, though, by the fact that the mountains continue to occupy a peripheral position, literally in terms of geography but also in regard to the centers of power and the interests of politicians. Several international organizations now intend to strengthen the voice of mountain people and move the plight of the Himalaya onto the agenda of global environmental programs.

The UNESCO Man and Biosphere Program (MAB) was one of the earliest international efforts to study Himalayan problems. Its work led to the formation in 1983 of the International Center for Integrated Mountain Development (ICIMOD). ICIMOD is based in Kathmandu, but it represents the interests of all the countries located in the Himalaya–Hindu Kush region. The International Mountain Society (IMS) also spearheaded early attempts to understand and resolve problems in the Himalaya as part of its global mountain agenda. The IMS, in collaboration with ICIMOD and other international organizations such as the International Union for the Conservation of Nature, organized a series of conferences, including the influential 1986 Mohonk Conference held in the United States, that led eventually to the writing of Chapter 13 of the United Nations Conference on Environment and Development (UNCED) in 1992. Chapter 13, known as the Mountain Agenda, promotes the global recognition of mountain areas as fragile ecosystems and advises special regulations for the international protection of mountains. Most recently, the study of land-cover change in the Himalaya has been included in the International Geosphere-Biosphere Program on Global Change.

Bhotiya shepherds in the high mountains of Nepal. Photo by P. P. Karan

An important outcome of these international efforts has been to encourage mountain countries to address problems that are unique to their highland areas. Nepal and Bhutan prepared country reports for the 1992 UNCED Earth Summit. India and Pakistan have organized numerous conferences on mountain issues. The G. B. Pant Institute of Himalayan Environment and Development, under the auspices of the Indian government, produced the 1992 Action Plan for the Himalaya. That document has been instrumental in guiding conservation and development in the Indian region. ICIMOD, meanwhile, has played a major role since 1983 in assessing nature-society conditions throughout the Himalayan region and in establishing a policy focus for sustainable development in the mountains.[10] These various collaborations have identified some of the key elements in the environmental future of the Himalaya.

Local conditions in the mountains continue to show alarming levels of human poverty and unsustainable uses of natural resources. A major thrust of regional development policies is toward resolving the dual problems of population growth and environmental management.

Despite numerous family planning and public health programs, fertility rates across the Himalaya region remain high, fueling rapid population growth. The population question may only be answered by a comprehensive effort linking fertility rates to the education of women, to new work opportunities, and to improved overall socioeconomic conditions. The success of such efforts, however, remains limited in most mountain places.

Nepal's trends in migration and urbanization concentrate population growth in certain localities but also relieve some pressures of population in the most densely settled parts of the country. Government programs of planned resettlement, such as the relocation of people from the Nepal hills to the tarai, are meant to regulate the movement of people and open up the frontiers to development. The unplanned migrations into the lowlands, however, have led to spontaneous land settlement with adverse social and environmental consequences.[11] Landlessness and poverty have deepened in the region, and immigrants from the mountains compete with native Tharu tribes for space and resources. The growth of towns in the mountains, meanwhile, has introduced a host of new problems to the region, relating to air, water, and soil contamination, as well as to the concentration of resource extraction around the town fringes.

The larger problem of environmental management is conceived in a broad sense as the proper use of resources. Since most people in the Himalaya are farmers, the nexus of much of the problem is in the linkage between farms and forests. The sustainability of mountain agriculture is vital for maintaining village households as well as for environmental quality. Farm-forest interaction is especially important insofar as it drives the use of biomass and the efficient recycling of organic matter. A great deal of attention is placed on how people use trees for energy and construction. A continued supply of wood is crucial for the mountain livelihoods; hence, much attention is given to maintaining a renewable timber supply. Current research on agroforestry, multipurpose trees, and seedling nurseries concentrates on tree-planting efforts and other strategies meant to reduce the demand for wood—for example, the design of more efficient stoves or houses.

Since the local farming systems also include livestock, the concerns about grazing, fodder, and pasture management are important. The grazing of the forest undergrowth by free roaming cattle and the intensive lopping of forests for fodder degrade the quality of existing forests across the range. Whereas forest areas may not be decreasing everywhere, the quality of canopy forests in many places is threatened. The practice of stall-feeding cattle reduces the impact of open grazing in the forests. Still, the condition of many highland pastures has steadily worsened in recent decades.

Degradation of range lands is linked to increases in the number of livestock and to inappropriate land-use practices. These are mainly domestic matters, but the degradation also may be attributed to the loss of herders' rights to pasturage, to the exploitation of mineral resources in the range lands, to limited veterinary or other support services, and to the encroachment of activities such as cash cropping, tourism and infrastructures onto the fragile grasslands. The condition of the range lands is especially critical because they contain some of the most fragile natural environments and occupy the headwaters of the major mountain rivers.

Indigenous Rights and Territory

Unlike the big development projects, which issue from distant political centers and promote national or regional wealth, sustainable-development programs are more localized and require the active participation of villagers. In the former case, the rights of people are often denied or ignored. In the latter case, however, the participation of rural people acknowledges their authority over land and local resources. Hence, in terms of indigenous rights and the preservation of traditional life in the Himalaya, the two development approaches are worlds apart.

The notion of most big development is that local people are backward and inward-looking and thus hinder progress and modernity, and big-development strategy is to replace the traditional ways with more modern ones. It is clear, however, that the traditional adaptations of mountain people have allowed their long-term survival in the Himalaya and represent sophisticated appraisals of the natural environment. Studies of village land use, for example, show how local environmental management is based upon a deep understanding of such complex ecological processes as soil erosion, forest regeneration, slope dynamics, and mass wasting.[12] The most innovative

Townsfolk gather at a weekly market in Gangtok, the capital of Sikkim. Photo by D. Zurick

sustainable-development programs build upon the base of traditional cultural knowledge, wedding that to new scientific approaches.

It is difficult for some of the development experts to accept indigenous knowledge because that wisdom appears to be unscientific. Consequently, the role of native peoples as stewards of the land may be circumvented in these programs. The preservation of forest groves for religious purposes, for example, may be dismissed as superstition, even though it constitutes an invaluable kind of forest protection that is found throughout the region. The rich knowledge of forest dwellers about plant and animal species may not be realized because it supports the traditional practices of tribal peoples rather than those of modern society. Yet it is clear from recent ethnobotany research that native forest people have enormous knowledge about the food and medicinal properties of native flora and fauna. The careful management and timing of water use in the Himalaya often is tied to deity worship and thus may be denied any validity by outsiders despite its proven worth. Moreover, much of local environmental knowledge is held by women, who have historically been denied participation in decision making and change. For these reasons, only recently has traditional knowledge in the Himalaya gained attention for its role in managing environmental resources.[13]

The authority that rests in the knowledge of the native people is tied to their territorial claims, which in turn are based on long-standing residence in a place. In a way, that gives native people inalienable rights to the land. The two elements of environmental knowledge and land rights are crucial for the survival of the Himalayan cultures, but they are too often ignored by the forces of modernity. The boundaries of national political space, of development projects, and even of parklands and protected areas reflect mainly outside interests; they usurp local claims to the land and the meanings given to it by indigenous peoples. We have seen earlier that this has been the case historically all across the Himalaya; it is reproduced today in countless modern local settings.

To imagine, though, that native people will quietly acquiesce to the appropriation of their land and resources is to discount the history of territorial conflicts in the mountains. Today, ongoing environmental struggles in such places as Garhwal, the Doon Valley, and numerous other localities where the large dams are planned or where commercial resource extractions are particularly rapacious show that the opposite is in fact the more likely case. The struggle for separate statehood among the mountain districts of Uttarkhand in northern Uttar Pradesh has turned some of these local struggles into a national Indian event.

In an interview for the *Indian Express* conducted at the height of Uttarkhand agitation, Sunderlal Bahuguna commented that "for a long time now, the local aspirations of people of Garhwal have been sacrificed for so called national interests. For centuries they have been neglected or insulted. But not anymore. The seeds of revolt are sown, the tide of Uttarkhand movement cannot be stopped."[14]

Uttarkhand separatism seeks to reclaim local authority over land and mountain resources previously lost to the power of the plains that emanates from the state capital in Lucknow. Because it covers such a vast territory, comprising all of the Garhwal and Kumaun districts, the Uttarkhand movement stands out from many of the smaller territorial struggles located in the mountains. But similar circumstances do exist elsewhere in the Himalaya; for example, in the Assam highlands, in Nepalese Gorkhaland in Sikkim, and in Kashmir.

Resolution of the territorial struggles in the Himalaya is crucial for the successful overall management of environmental resources there, and incorporation of traditional knowledge is the basis for engaging the participation of local people in the sustainable mountain development projects. That is especially true for those efforts that seek to preserve the rich diversity of habitats in the mountains. It is now widely acknowledged that biodiversity in the Himalaya is analogous to cultural diversity. The roles of local knowledge and indigenous rights, which contribute to the cultural complexity of the mountains, assume also a central place in the management of the region's environmental diversity.[15]

Reflections on a Mountain Range

The Himalaya today is a dramatically changed place, where modernity has been telescoped into a very few decades. The image of a mountain fastness, a Shangri La, which is naturally evoked in the landscapes of the Himalaya or which has been purposely foisted upon it to serve the purposes of the Western imagination, has given way to a new world identity. The transformations that best describe the contemporary Himalaya are compelled by the circumstances of its own history, by the struggles of its native peoples, and by the designs of national development. Life and land in the mountains today connect more fully than ever before the remote highlands and the affairs of the wider world. This requires a new image of the Himalaya, as a place not separate from the rest of the planet but an integral part of it.

The widespread attention given to Himalayan water resources, especially in terms of hydropower development and the irrigation of agriculture in the plains, has lent a renewed vigor to international assessments of the Himalayan environment. The possibilities of cross-border trade are excited not only by the fiscal policies of the mountain countries but also by the placement of new roadways, airline connections, and other transportation infrastructures that connect the diverse geographies in the mountains. The new trends of transboundary tourism and the establishment of international mountain parklands in northern Pakistan, Nepal, and Tibet instill ever-widening social, economic, and natural interactions across the region's political borders.

As the scale of environmental policies begins to match the realities of ecological systems, the need for international cooperation and transborder environmental management will increase. That already is evident in a positive way in the management of some watersheds that

The cultural landscape in Bhutan shows a mix of religious buildings, small settlements, roads, and farms. Photo by P. P. Karan

cross international borders, of some protected areas that adjoin national territories—for examples, the combined Sagarmatha National
Park–Qomolangma Nature Preserve in Nepal and Tibet and the Manas
Park straddling the Bhutan-India border—and in the establishment
in 1990 of a two-kilometer-wide green belt along the entire southern
border of Bhutan.

The transboundary dimensions of environmental change are apparent also in some potentially damaging trends in the region. The
cross-border smuggling of medicinal plants and wildlife products, to
give one instance, is widespread in the Himalaya and is made possible
by inadequate legislation or lack of government surveillance. As in-
dustrialization proceeds in the mountains, the problems of atmos-
pheric pollution will increase. As an extreme example, there is some
concern already that nuclear power plants under construction in
India along the Himalayan foothills may produce radioactive fallout
that could contaminate the glaciers and snowfields of the Great Himalaya.

The dominant environmental model of the past few decades attempts to link the Himalaya with the southern Asian plains in a grand
scheme of highland-lowland interaction. According to that model, the
poor management of the mountain slope lands, especially the forested
areas, adversely affects the status of environmental conditions in
places as far away as the Bay of Bengal. The scale of this purported linkage is massive, covering as it does the entire Himalayan range, much
of the Tibetan plateau, and the huge lowland basins of the Indus,
Ganges, and Brahmaputra rivers. Great uncertainty attends this particular scenario, especially the linkage between vegetation change in
the highlands and river floods in the lowlands, but scientific studies
clearly show the alarming status of environment and social conditions
in many Himalaya places. We have seen, for example, places where resource intensification and land degradation in the mountains is driven by poverty, by a region's imperial history, by the demands of a burgeoning subsistence economy, and by the incessant commercial
interests of modern development. Moreover, the mix of these factors,
and their consequences, varies widely across the mountain terrain.
The natural diversity of the Himalaya is matched by equally complex
social arrangements. Combined, they determine the scope of the environmental degradation, as well as the prospects for its resolution.

The highly variable conditions in the Himalaya reject the application of a single environmental model, but that does not mean the situation is not bad. In fact, the trends of habitat destruction, soil erosion, declining farmland production, and human poverty are so alarming and widespread as to give the range its much deserved international concern. This is reflected in the various global environmental initiatives, including the Mountain Agenda of the United Nations, that are centered on the Himalaya. Morever, local problems are so widely replicated across the Himalaya that their cumulative impacts may well have global consequences.

The marginal conditions which modernity engenders among native cultures contribute in a deep way to the greater vulnerability of people attempting to survive in conditions of environmental instability. The traditional cultural adaptations, designed to ride the fluctuating risks inherent in the natural conditions of the mountains, may not apply equally well to the vagaries of the rest of the world or to the ecological shifts brought about by the many new developments. That is especially true where local authority over land and resources is diminished by political or economic events. Much of the current instability in the Himalaya is due to the fact that the forces of change that affect local communities have their genesis in distant societies and therefore lie outside the experiences of mountain people. The new conditions of modern life determine much of the course of change in the mountains, but they have not yet totally usurped the cultural resiliency that has built up among resident cultures over the centuries. The fact that mountain environments, despite their appearances of immutability, are fragile places requiring unique environmental management compels close attention to the sustainable practices of the past as well as to modern innovations. The uncertainty of life in the Himalaya today is accelerated when this attention is not forthcoming.

District key

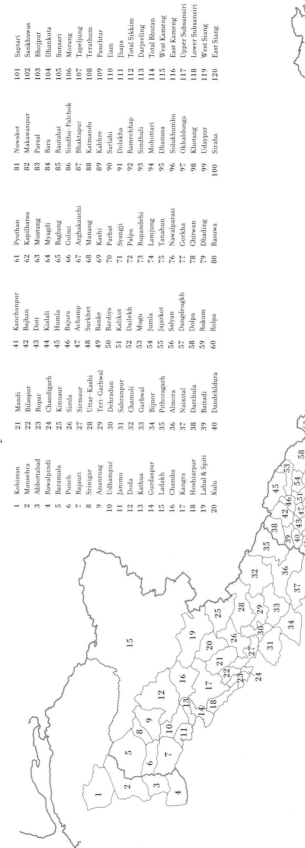

1 Kohistan	41 Kanchanpur	61 Pyuthan	81 Nuwakot	101 Saptari
2 Mansehra	42 Bajhan	62 Kapilbatsu	82 Makawanpur	102 Sankhuwas
3 Abbottabad	43 Doti	63 Mustang	83 Parsal	103 Bhojpur
4 Rawalpindi	44 Kialali	64 Myagdi	84 Bara	104 Dhankuta
5 Baramula	45 Humla	65 Baglung	85 Rautahat	105 Sunsari
6 Punch	46 Bajura	66 Gulmi	86 Sindhu-Palchok	106 Morang
7 Rajauri	47 Achamp	67 Arghakanchi	87 Bhaktapur	107 Tapeljung
8 Srinagar	48 Surkhet	68 Manang	88 Katmandu	108 Terathum
9 Anantnag	49 Banke	69 Kashi	89 Kabhre	109 Panchtar
10 Udhampur	50 Bardiya	70 Parbat	90 Sarlahi	110 Ilam
11 Jammu	51 Kalikot	71 Syangji	91 Dolakha	111 Jhapa
12 Doda	52 Dailekh	72 Palpa	92 Ramechhap	112 Total Sikkim
13 Kathua	53 Mugu	73 Rupandehi	93 Sindhuli	113 Darjeeling
14 Gurdaspur	54 Jumla	74 Lamjung	94 Mohottari	114 Total Bhutan
15 Ladakh	55 Jajarkot	75 Tanahun	95 Dhanusa	115 West Kameng
16 Chamba	56 Salyan	76 Nawalparasi	96 Solukhumbu	116 East Kameng
17 Kangra	57 Dangdeugkh	77 Gorkha	97 Okhaldunga	117 Upper Subsansiri
18 Hoshiarpur	58 Dolpa	78 Chitwan	98 Khotang	118 Lower Subsansiri
19 Lahul & Spiti	59 Rukum	79 Dhading	99 Udaypur	119 West Siang
20 Kulu	60 Rolpa	80 Rasuwa	100 Siraha	120 East Siang

21 Mandi	31 Saharanpur
22 Bilaspur	32 Chamoli
23 Ropar	33 Garhwal
24 Chandigarh	34 Bijnor
25 Kinnaur	35 Pithoragarh
26 Simla	36 Almora
27 Sirmaur	37 Nainital
28 Uttar-Kashi	38 Darchula
29 Teri-Garhwal	39 Baitadi
30 Dehradun	40 Dandehldura

Appendix

THE FOLLOWING TABLES provide detailed numerical information on all the districts in the Himalayan countries. The district numbers correspond to the district key map. The tables are meant for those readers who wish to examine in more detail the human-environment trends in specific districts or to compare trends across the region. The numerical information contained in the tables provides the statistical bases for many of the maps that appear in the text. Because the data in the tables are compiled from national census reports and other government records, their accuracy reflects the level of precision of the Himalayan countries' census taking and record keeping. Although specific numbers may be questionable, especially for the historical periods, overall this data set represents the only existing compilation of the Himalayan figures and provides a valuable basis for comparing trends across the mountain range.

Table A1 Total Population (1890–1990)

District	1890	1900	1910	1920	1930	1940	1950	1960	1970	1980	1990
1							147,403	151,718	195,580	465,237	
2	165,312						266,613	353,000	518,017	1,067,000	
3	175,735						337,378	373,903	551,144	1,169,432	
4	887,194	930,535	547,827	569,224	634,357	785,231	907,794	1,084,260	1,747,685	2,121,000	
5			460,515	502,216	559,826	612,428		604,659	775,724	670,142	
6	268,608	304,448	334,393	351,781	387,384	421,828		326,061	170,787		224,197
7					128,369	140,844				302,500	
8	118,960	122,618			771,943				827,697	708,328	
9	224,558		639,481	688,761	771,943	851,606		654,368	832,280	656,351	
10	150,707	284,048	215,725	218,261	273,668	294,217		254,061	338,846	453,636	
11	1,439,543	1,521,307	326,691	334,834	375,240	431,362		516,932	731,743	943,395	
12								268,403	342,220	425,262	
13	151,518	154,213	151,802	154,209	161,232	177,672		207,430	274,671	369,123	
14	943,922	940,334	836,771	852,192	970,898	1,153,511	851,294	987,994	1,229,249	1,513,435	1,765,834
15	28,274	165,992	186,656	183,476	192,138	195,431		88,651	105,291	68,380	
16	124,032	127,834	135,873	141,867	146,870	168,908	176,050	210,579	255,233	311,147	391,047
17	763,030	768,124	770,386	766,065	801,312	899,377	936,042	1,062,518	1,327,211	990,758	1,149,744
18	1,011,659	989,782	918,569	927,419	1,032,187	1,170,323	1,091,986	1,233,493	1,052,153	1,243,807	1,430,210
19	5,982							20,453	23,538	32,100	31,294
20	55,100								192,371	238,734	301,729
21	166,923	174,045	181,110	185,408	207,465	232,593	310,626	384,259	515,180	644,827	768,446
22	91,760	90,873	93,107	98,000	100,994	110,336	126,099	158,806	194,786	247,368	291,388
23	146,816								545,005	716,662	905,629
24									257,251	451,610	640,725
25								40,980	49,835	59,547	70,931

26	44,642	40,351	39,320	45,327	36,786	38,576	46,150	112,653	217,129	510,932	614,404
27	124,134	135,687	138,520	140,448	148,568	156,028	166,077	197,551	245,033	306,952	379,801
28	69,209	69,209	77,429	81,958	89,978	102,280	106,058	122,836	147,805	190,948	239,772
29	241,242	268,885	300,819	318,414	349,573	397,369	412,047	347,736	397,385	497,710	580,352
30	168,135	178,195	205,075	212,243	230,247	266,244	211,041	429,014	577,306	761,668	1,025,700
31	1,001,280	1,045,230	986,359	937,471	1,043,920	1,179,643	1,353,636	1,615,478	2,054,834	2,673,561	2,309,495
32	145,670	162,703	164,584	181,103	204,248	216,972	253,137	292,571		364,346	454,667
33	407,818	429,900	480,167	485,186	533,885	602,115	639,625	482,327	553,028	637,877	682,165
34	764,070	779,951	806,202	740,182	835,469	910,223	984,196	1,190,987	1,490,185	1,939,261	2,454,989
35	910,895	912,848	899,973	855,130	906,233	1,041,024	1,110,734	263,579	313,747	489,267	566,148
36	411,501	465,893	525,104	530,338	583,302	687,286	772,896	633,407	750,038	757,373	836,134
37	368,312	311,237	323,519	276,875	277,286	291,861	335,414	574,320	790,080	1,136,523	1,557,415
38									68,868	90,218	101,614
39						75,457	133,231	158,612	128,696	180,136	200,229
40						37,228	69,619	82,709	94,743	86,853	104,449
41						11,767	17,745	18,869	68,863	168,979	258,508
42									108,623	124,006	139,178
43	216,489					207,562	254,314	282,637	166,070	153,132	167,469
44	46,816					37,435	76,606	89,795	128,877	257,905	420,035
45									29,524	20,303	34,640
46									61,342	74,649	92,083
47	17,327						152,520	151,365	131,612	185,212	197,888
48						26,963	34,939		104,933	166,166	225,296
49						72,969	82,449	95,096	125,709	205,323	284,430
50						43,842	58,389	67,834	101,793	199,044	289,840
51									10,017	87,636	88,781
52	84,173					183,670	185,854	210,360	156,072	166,527	187,820
53									25,718	43,705	36,445

Table A1 *(continued)*

District	1890	1900	1910	1920	1930	1940	1950	1960	1970	1980	1990
54				89,022		118,896	166,360	184,505	122,753	68,790	76,305
55									85,654	99,315	114,267
56						306,922	375,507	415,524	142,457	152,063	182,145
57				292,733		382,263	471,314	98,765	167,820	266,393	
58									19,110	22,033	25,075
59									96,243	132,432	155,017
60									162,955	168,166	179,904
61				122,063		173,228	198,138	212,481	137,338	157,669	173,893
62									205,216	270,045	372,205
63									26,944	12,930	14,319
64									57,946	96,904	100,918
65							218,936	228,769	172,729	215,228	232,737
66							272,805	301,730	227,946	238,113	265,707
67									130,212	157,304	180,884
68									7,436	7,021	5,369
69						94,358	111,747	127,515	151,749	221,272	293,009
70									118,689	132,192	143,753
71				183,417		256,941	317,785	339,634	268,606	228,685	292,014
72						104,720	147,551	172,307	212,633	214,442	236,238
73									243,386	379,166	507,689
74						107,543	125,329	130,935	140,222	152,720	
75				82,160		72,878	113,677	127,642	159,939	223,438	266,766
92									146,548	309,332	435,256
77							135,975	151,264	178,265	230,714	251,750
88				20,520		26,239	42,724	67,882	183,644	259,571	355,298

79		119,109	175,282	203,039	236,276	243,401	278,488
80		120,019			17,517	30,241	36,768
81			147,496	162,981	172,718	202,976	245,645
82					163,764	243,411	315,588
83		104,820	114,107	137,624	202,123	284,338	371,533
84		178,624	210,007	250,054	233,401	318,957	413,294
85		168,226	189,223	218,661	320,093	332,526	412,921
86		134,782	174,630	186,492	206,384	232,326	260,972
87			83,460	89,822	110,157	159,767	173,097
88			327,535	380,168	508,754	606,547	927,079
89		114,005	187,685	216,440	245,165	307,150	324,819
90		129,944	134,340	162,979	175,543	398,766	490,390
91			125,384	132,112	130,022	150,576	173,836
92			116,577	123,357	157,349	161,445	188,814
93			100,663	104,237	147,417	183,705	222,946
94			418,436	488,218	324,827	361,054	440,774
95					330,601	432,569	54,975
96					105,324	88,245	97,253
97			145,147	164,429	122,872	137,640	140,436
98					163,297	212,571	216,820
99	48,913	39,486	88,565	89,429	112,622	159,805	218,889
100	377,855	163,894	176,747	210,529	302,304	373,358	460,122
101		363,941	431,573		312,565	379,055	464,500
102					114,313	129,414	141,771
103			252,660	194,506	192,689	198,301	
104			514,216		107,649	129,781	17,155
105					223,434	344,558	464,767
106	211,308	241,474	309,204		301,507	534,694	676,417

Table A₁ (continued)

District	1890	1900	1910	1920	1930	1940	1950	1960	1970	1980	1990
107									84,760	120,780	120,072
108						276,461	375,246	422,952	119,307	92,454	102,952
109									145,809	153,746	175,821
110				87,475		91,362	112,848	124,525	139,538	178,356	229,429
111						69,045	80,158	119,700	247,698	479,743	594,100
112	30,458	59,014	87,920	81,721	109,808	121,520	137,725	162,189	209,843	316,385	403,612
113	223,314	249,117	265,550	282,748	319,635	376,369	445,260	624,640	781,777	1,024,269	1,335,618
114								9,630,614		1,162,000	
115										63,302	56,402
116										42,736	50,238
117										39,410	112,650
118										112,650	154,591
119										74,164	89,778
120										70,451	99,985

Table A2 Arithmetic Population Density: 1970 and 1990 (persons/hectare)

District	Hectares		Population Density	
	1970	1990	1970	1990
1		758,071		
2		595,698		
3		36,499		
4	204,092	528,609	8.56	
5	745,800	458,800	1.04	
6	114,000	167,400	1.50	1.34
7	253,000	263,000		
8	128,000	222,800	6.47	
9	538,200	398,400	1.55	
10	454,900	455,000	0.74	
11	320,000	309,700	2.29	
12	1,169,100	1,169,100	0.29	
13	265,100	265,100	1.04	
14	356,000	356,200	3.45	4.96
15	61,000	8,266,500	1.73	
16	819,500	652,800	0.31	0.60
17	839,700	573,900	1.58	2.00
18	388,300	388,100	2.71	3.69
19	1,201,500	1,383,500	0.02	0.02
20	543,500	550,300	0.35	0.55
21	401,800	395,500	1.28	1.94
22	116,700	116,700	1.67	2.50
23	208,500	208,500	2.61	4.34
24	11,400	11,400	22.57	56.20
25	655,300	640,100	0.08	0.11
26	141,600	513,100	1.53	1.20
27	282,500	282,500	0.87	1.34
28	801,600	801,600	0.18	0.30
29	442,100	442,100	0.90	1.31
30	308,800	308,800	1.87	3.32
31	552,600	368,900	3.72	6.26
32	912,500	912,600	0.32	0.50
33	544,000	543,800	1.02	1.25
34	485,200	456,100	3.07	5.38
35	721,700	885,600	0.43	0.64
36	702,300	538,500	1.07	1.55

Table A2 (continued)

District	Hectares		Population Density	
	1970	1990	1970	1990
37	679,200	679,400	1.16	2.29
38	232,200	232,200	0.30	0.44
39	151,900	151,900	0.85	1.32
40	153,800	153,800	0.62	0.68
41	161,000	161,000	0.43	1.61
42	342,200	342,200	0.32	0.41
43	202,500	202,500	0.82	0.83
44	323,500	323,500	0.40	1.30
45	565,500	565,500	0.05	0.06
46	218,800	218,800	0.28	0.42
47	168,000	168,000	0.78	1.18
48	245,100	245,100	0.43	0.92
49	233,700	233,700	0.54	1.22
50	202,500	202,500	0.50	1.43
51	174,100	174,100	0.06	0.51
52	150,200	150,200	1.04	1.25
53	353,500	353,500	0.07	0.10
54	253,100	253,100	0.48	0.30
55	223,000	223,000	0.38	0.51
56	146,200	146,200	0.97	1.25
57	295,500	295,500	0.57	
58	788,900	788,900	0.02	0.03
59	287,700	287,700	0.33	0.54
60	187,900	187,900	0.87	0.96
61	130,900	130,900	1.05	1.33
62	173,800	173,800	1.18	2.14
63	357,300	357,300	0.08	0.04
64	229,700	229,700	0.25	0.44
65	178,400	178,400	0.97	1.30
66	114,900	114,900	1.98	2.31
67	119,300	119,300	1.09	1.52
68	224,600	224,600	0.03	0.02
69	201,700	201,700	0.75	1.45
70	49,400	49,400	2.40	2.91
71	116,400	116,400	2.31	2.51
72	137,300	137,300	1.55	1.72
73	136,000	136,000	1.79	3.73

Table A2 (continued)

District	Hectares		Population Density	
	1970	1990	1970	1990
74	169,200	169,200	0.83	
75	154,600	154,600	1.03	1.73
76	216,200	216,200	0.68	2.01
77	361,000	361,000	0.49	0.70
78	221,800	221,800	0.83	1.60
79	192,600	192,600	1.23	1.45
80	154,400	154,400	0.11	0.24
81	112,100	112,100	1.54	2.19
82	242,600	242,600	0.68	1.30
83	135,300	135,300	1.49	2.75
84	119,000	119,000	1.96	3.47
85	112,600	112,600	2.84	3.67
86	254,200	254,200	0.81	1.03
87	11,900	11,900	9.26	14.55
88	78,000	78,000	6.52	11.89
89	139,600	139,600	1.76	2.33
90	125,900	125,900	1.39	3.90
91	219,100	219,100	0.59	0.79
92	154,600	154,600	1.02	1.22
93	249,100	249,100	0.59	0.90
94	100,200	100,200	3.24	4.40
95	118,000	118,000	2.80	0.47
96	331,200	331,200	0.32	0.29
97	107,400	107,400	1.14	1.31
98	159,100	159,100	1.03	1.36
99	206,300	206,300	0.55	1.06
100	118,800	118,800	2.54	3.87
101	136,300	136,300	2.29	3.41
102	348,000	348,000	0.33	0.41
103	150,700	150,700	1.28	
104	89,100	89,100	1.21	0.19
105	125,700	125,700	1.78	3.70
106	185,500	185,500	1.63	3.65
107	364,600	364,600	0.23	0.33
108	67,900	67,900	1.76	1.52
109	124,100	124,100	1.17	1.42
110	170,300	170,300	0.82	1.35

Table A2 (continued)

District	Hectares		Population Density	
	1970	1990	1970	1990
111	160,600	160,600	1.54	3.70
112	729,900	709,600	0.29	0.57
113	307,500	314,900	2.54	4.24
114		4,650,000		
115		979,400		0.06
116		413,400		0.12
117		703,200		0.07
118		1,301,000		0.12
119		1,200,600		0.07
120		651,200		0.15

Table A3 Total Forest Area (hectares) (1890–1990)

District	1890	1900	1910	1920	1930	1940	1950	1960	1970	1980	1990
1								206,798		200,000	
2										61,283	
3										138,874	
4		132,249			66,882				103,600	87,971	
5					335,000	287,200	391,600	125,100	174,800	129,600	250,000
6						10,968		138,543	32,729	34,306	133,100
7							219,174		103,270	92,521	
8					652	735	2,084	1,254	2,051	1,274	
9							74,800	115,900	116,000	47,800	218,600
10				710,153	726,242	720,618	472,684	192,078	191,167	188,919	447,000
11				145,556	104,563	97,939	15,444	48,328	48,243	48,266	74,400
12							223,951	224,340	220,620	219,746	
13					166,888	171,673	171,679	72,294	69,708		107,400
14		6,146		11,866	5,910			12,000	16,000	15,466	17,100
15					113,789	2,671	9,309				2,000
16						31,900	118,839	801,145	95,510	252,354	212,400
17					665,951		662,535		201,930	188,441	175,500
18		52,810			52,934		35,813	35,813	76,472	93,723	71,900
19									12,350	43,534	1,900
20										194,800	204,400
21				50,457	51,130		94,876	184,685	148,140	150,250	130,900
22					13,526	5,640		12,027	11,890	11,474	15,700
23								43,000	40,000	40,000	34,400
24										500	500
25										63,300	62,900

Table A3 (continued)

District	1890	1900	1910	1920	1930	1940	1950	1960	1970	1980	1990
26		3,676			4,511					48,879	242,500
27					159,110	59,100	129,753	122,575	46,759	48,260	101,900
28					1,365,237			1,623,818	704,476	710,276	310,300
29						1,775,987	1,775,986		269,545	397,201	256,200
30		46,500			639,283		278,651	384,203	167,378	226,086	157,000
31					186,969		186,969	190,564	76,388	77,343	70,700
32							2,056,707	2,054,707	521,828	526,936	315,100
33					3,301,000				455,612	314,900	317,800
34					58,752		57,016	171,781	58,690	70,847	55,700
35									237,915	330,288	298,300
36					1,235,127		690,151	690,101		394,443	253,700
37			89,183		872,563		1,147,976	874,819	66,283	402,558	360,300
38								104,000	50,000	79,537	
39								129,600	122,500	78,720	
40								107,200	125,000	115,891	
41								127,950	127,950	100,062	53,371
42								136,000	70,000	113,179	
43								126,400	157,500	144,689	
44								184,278	184,278	231,092	156,631
45								100,800	85,000	74,783	
46								107,200	40,000	98,646	
47								110,400	122,500	88,097	
48								166,400	140,000	177,854	
49								164,746	164,746	167,198	121,957
50								130,277	130,277	127,682	27,906

51	108,800	135,000	106,055	
52	56,000	68,648	78,136	
53	99,200	105,000	111,096	
54	102,400	40,000	104,570	
55	116,800	90,000	135,615	
56	86,400	87,433	70,597	
57	97,736	97,736	197,272	139,726
58	161,600	90,000	63,875	
59	160,000	168,595	136,452	
60	83,200	122,351	94,096	
61	62,400	17,500	72,694	
62	94,260	94,260	77,838	77,401
63	40,000	35,700	184,882	
64	80,000	95,741	84,452	
65	81,600	125,459	98,045	
66	68,800	64,483	40,663	
67	52,800	45,000	73,133	
68	38,400	57,500	19,165	
69	44,800	20,000	89,943	
70		27,465	19,997	
71		59,104	31,690	
72	19,200	110,000	71,172	
73	44,800	75,000	39,332	29,572
74	60,800	40,000	84,316	
75	32,000	37,500	8,562	
76	131,200	130,000	114,899	85,863
77	81,600	97,500	112,534	
78	179,200	26,272	142,421	59,109

Table A3 (continued)

District	1890	1900	1910	1920	1930	1940	1950	1960	1970	1980	1990
79								46,400	75,000	92,855	
80								65,600	59,338	52,290	
81								28,800	44,963	49,654	
82								150,400	130,000	167,453	
83								59,200	67,500	77,123	25,296
84								51,200	70,000	52,558	43,551
85								20,800	40,000	30,752	29,402
86								92,800	80,000	126,543	
87								1,600	4,508	1,947	
88								22,400	50,294	34,479	
89								40,000	72,002	73,800	
90								67,200	35,000	32,244	25,557
91								88,000	32,500	94,477	
92								35,200	74,156	66,151	
93								113,600	95,000	178,129	
94								49,600	30,000	24,456	21,945
95								51,200	32,500	30,399	24,737
96								96,000	80,000	105,329	
97								19,200	61,031	47,346	
98								30,400	76,766	79,553	
99								110,400	87,500	138,916	322,800
100								19,200	7,500	28,154	21,811
101								22,200	5,000	341,892	3,300
102								184,000	15,000	180,580	
103								51,200	83,422	77,887	

104				18,400	52,437	36,382	
105				24,000	32,500	23,204	15,009
106				100,800	27,000	54,691	55,488
107				180,800	95,000	139,167	
108				12,800	25,185	24,627	
109					42,500	57,706	
110				75,200	65,000	95,918	
111					11,500	17,107	15,000
112						262,200	311,900
113		117,160	117,240	118,300	128,500		145,500
114	278,890					2,825,000	2,646,300
115							
116							
117							
118							
119							
120							

287,347

Table A4 Population–Forest Area (persons/hectare of forest)

District	Forest Area (hectares)		Population–Forest Density	
	1970	1990	1970	1990
1				
2				
3				
4	103,600		16.87	
5	174,800	250,000	4.44	
6	32,729	133,100	5.22	1.68
7	103,270			
8	2,051		403.56	
9	116,000	218,600	7.17	
10	191,167	447,000	1.77	
11	48,243	74,400	15.17	
12	220,620		1.55	
13	69,708	107,400	3.94	
14	16,000	17,100	76.83	103.27
15		2,000		
16	95,510	212,400	2.67	1.84
17	201,930	175,500	6.57	6.55
18	76,472	71,900	13.76	19.89
19	12,350	1,900	1.91	16.47
20		204,400		1.48
21	148,140	130,900	3.48	5.87
22	11,890	15,700	16.38	18.56
23	40,000	34,400	13.63	26.33
24		500		1,281.45
25		62,900		1.13
26	48,620	242,500	4.47	2.53
27	46,759	101,900	5.24	3.73
28	704,476	310,300	0.21	0.77
29	269,545	256,200	1.47	2.27
30	167,378	157,000	3.45	6.53
31	76,388	70,700	26.90	32.67
32	521,828	315,100	0.56	1.44
33	455,612	317,800	1.21	2.15
34	58,690	55,700	25.39	44.08
35	237,915	298,300	1.32	1.90
36		253,700		3.30
37	66,283	360,300	11.92	4.32

District	Forest Area (hectares)		Population-Forest Density	
	1970	1990	1970	1990
38	50,000		1.38	
39	122,500		1.05	
40	125,000		0.76	
41	127,950	53,371	0.54	4.84
42	70,000		1.55	
43	157,500		1.05	
44	184,278	156,631	0.70	2.68
45	85,000		0.35	
46	40,000		1.53	
47	122,500		1.07	
48	140,000		0.75	
49	164,746	121,957	0.76	2.33
50	130,277	27,906	0.78	10.39
51	135,000		0.07	
52	68,648		2.27	
53	105,000		0.24	
54	40,000		3.07	
55	90,000		0.95	
56	87,433		1.63	
57	97,736	139,726	1.72	
58	90,000		0.21	
59	168,595		0.57	
60	122,351		1.33	
61	17,500		7.85	
62	94,260	67,401	2.18	5.52
63	35,700		0.75	
64	95,741		0.61	
65	125,459		1.38	
66	64,483		3.53	
67	45,000		2.89	
68	57,500		0.13	
69	20,000		7.59	
70	27,465		4.32	
71	59,104		4.54	
72	110,000		1.93	
73	75,000	29,572	3.25	17.17
74	40,000		3.51	

District	Forest Area (hectares)		Population-Forest Density	
	1970	1990	1970	1990
75	37,500		4.27	
76	130,000	85,863	1.13	5.07
77	97,500		1.83	
78	26,272	59,109	6.99	6.01
79	75,000		3.15	
80	59,338		0.30	
81	44,963		3.84	
82	130,000		1.26	
83	67,500	25,296	2.99	14.69
84	70,000	43,551	3.33	9.49
85	40,000	29,402	8.00	14.04
86	80,000		2.58	
87	4,508		24.44	
88	50,294		10.12	
89	72,002		3.40	
90	35,000	25,557	5.02	19.19
91	32,500		4.00	
92	74,156		2.12	
93	95,000		1.55	
94	30,000	21,945	10.83	20.09
95	32,500	24,737	10.17	2.22
96	80,000		1.32	
97	61,031		2.01	
98	76,766		2.13	
99	87,500	322,800	1.29	0.68
100	7,500	21,811	40.31	21.10
101	5,000	3,300	62.51	140.76
102	15,000		7.62	
103	83,422		2.31	
104	52,437		2.05	
105	32,500	15,009	6.87	30.97
106	27,000	55,488	11.17	12.19
107	95,000		0.89	
108	25,185		4.74	
109	42,500		3.43	
110	65,000		2.15	
111	11,500	15,000	21.54	39.61

District	Forest Area (hectares)		Population-Forest Density	
	1970	1990	1970	1990
112		311,900		1.29
113	128,500	145,500	6.08	9.18
114		2,646,300		
115				
116				
117				
118				
119				
120				

Table A5 Total Cropped Area (hectares) (1890–1990)

District	1890	1900	1910	1920	1930	1940	1950	1960	1970	1980	1990
1											
2										97,962	
3											
4	199,898	190,967	184,146	222,077	228,504		588,355	1,025,962			
5					389,158	398,858	287,486	13,545	131,210	89,349	
6						183,675		67,584	28,847	36,788	
7							75,735		64,047	76,518	
8					290,034	400,611	186,370	78,678	71,930	30,492	
9							312,570	135,777	136,206	86,308	
10				143,631	138,940	163,548	100,016	84,568	84,552	101,184	
11				248,021	254,601	119,224	118,980	387,421	171,746	184,952	
12							51,927	60,260	64,047	67,710	
13					131,777	157,813	166,536	81,828	89,315		
14	346,971	349,006	376,756	355,575	361,650		284,677	318,000	363,000	418,410	
15				79,428	76,052	77,410	60,109	16,983	17,000	9,630	
16							159,736	142,141	60,410	64,117	
17	253,305	265,548	269,169	274,269	273,649		316,174		196,340	193,227	
18	340,915	314,381	293,961	293,059	296,029		354,485	309,000	347,000	379,358	
19									2,860	3,144	
20									51,100	52,127	
21			58,091	57,415	56,610		284,887	323,488	136,851	159,214	
22					35,051			51,693	55,170	59,395	
23						2,149,189		174,000	190,000	194,000	
24										4,400	
25								26,154	11,220	10,405	

26	5,002				4,270		570	38,417	108,420	108,414	
27		4,749	5,154	4,978		92,190	159,853	168,596	72,507	71,310	
28								191,195	41,906	53,136	
29					283,083	220,809	188,126		128,566	123,607	
30		36,017			102,975		163,877	195,465	81,979	88,541	
31		347,749	328,955		181,286		1,149,721		1,276,002	553,250	601,943
32								962,486	86,595	72,424	
33					260,000		962,258		172,181	152,620	
34					649,139		831,763	925,932	419,175	471,640	
35		512,689							110,701	123,815	
36					295,000		505,026	514,122	261,805	188,964	
37		170,374			247,960		333,548	630,787	51,965	344,615	
38								4,800		24,964	
39								17,600		50,503	
40								14,400		27,233	
41								8,000		47,370	
42								14,400		41,715	
43								22,400		49,368	
44								32,000	653		
45								3,200		9,298	
46								12,800		23,345	
47								20,800		59,592	
48								25,600		52,200	
49								44,800		58,976	
50								30,400		69,963	
51								4,800		23,748	
52								17,600		52,721	
53								19,200		18,794	

Table A5 (continued)

District	1890	1900	1910	1920	1930	1940	1950	1960	1970	1980	1990
54								36,800		24,057	
55								91,200		35,222	
56								6,400		52,674	
57								56,000		80,784	
58								60,800		9,340	
59								43,200		45,300	
60								52,800		59,855	
61								4,800		43,041	
62								36,800		93,855	
63								36,800		5,468	
64								14,400		30,012	
65								30,400		52,343	
66								1,600		54,178	
67								56,000		39,947	
68								36,800		1,076	
69								83,200		53,004	
70								86,400		28,606	
71								67,200		60,017	
72								40,000		57,172	
73								104,000		97,906	
74								43,200		43,871	
75								86,400		66,100	
76								22,400		70,143	
77								94,400		64,538	
78								1,440,008		55,279	

73,346	51,200	79
10,381	52,800	80
60,151	49,600	81
59,657	60,800	82
54,732	49,600	83
69,963	81,600	84
64,599	8,000	85
63,905	57,600	86
9,680	75,200	87
42,263	41,600	88
61,599	60,800	89
84,736	94,400	90
44,871	104,000	91
59,179	48,000	92
58,846	57,600	93
69,323	51,200	94
83,181	59,200	95
33,313	100,800	96
48,838	73,600	97
71,322	75,200	98
52,393	96,000	99
38,355	40,000	100
86,385	78,400	101
49,468	40,000	102
65,733	78,400	103
46,816	40,000	104
81,944	52,800	105
123,070	52,800	106

Table A5 (continued)

District	1890	1900	1910	1920	1930	1940	1950	1960	1970	1980	1990
107								59,200		71,748	
108								51,200		41,392	
109								0		65,548	
110								96,000		70,954	
111										126,632	
112										91,300	
113			174,600		63,901		240,000	290,700	45,838		
114										112,416	131,000
115											
116											
117											
118											
119											
120											

Table A6 Population–Cropped Area (persons/hectare)

District	Cropped Area (hectares)		Population–Cropped-Area Density	
	1960	1980	1960	1980
1				
2		97,962	10.89	
3				
4	1,025,962			1.06
5	13,545	89,349	7.50	44.64
6	67,584	36,788	0.00	4.82
7		76,518	3.95	
8	78,678	30,492	23.23	
9	135,777	86,308	7.60	4.82
10	84,568	101,184	4.48	3.00
11	387,421	184,952	5.10	1.33
12	60,260	67,710	6.28	4.45
13	81,828			2.53
14	318,000	418,410	3.62	3.11
15	16,983	9,630	7.10	5.22
16	142,141	64,117	4.85	1.48
17		193,227	5.13	
18	309,000	379,358	3.28	3.99
19		3,144	10.21	
20		52,127	4.58	
21	323,488	159,214	4.05	1.19
22	51,693	59,395	4.16	3.07
23	174,000	194,000	3.69	
24		4,400	102.64	
25	26,154	10,405	5.72	1.57
26	38,417	108,414	4.71	2.93
27	168,596	71,310	4.30	1.17
28	191,195	53,136	3.59	0.64
29		123,607	4.03	
30	195,465	88,541	8.60	2.19
31		553,250	4.83	
32	962,486	72,424	5.03	0.26
33		152,620	4.18	
34	925,932	471,640	4.11	1.29
35		123,815	3.95	
36	514,122	188,964	4.01	1.23
37	630,787	344,615	3.30	0.91

District	Cropped Area (hectares)		Population—Cropped-Area Density	
	1960	1980	1960	1980
38	4,800	24,964	3.61	
39	17,600	50,503	3.57	9.01
40	14,400	27,233	3.19	5.74
41	8,000	47,370	3.57	2.36
42	14,400	41,715	2.97	
43	22,400	49,368	3.10	12.62
44	32,000			2.81
45	3,200	9,298	2.18	
46	12,800	23,345	3.20	
47	20,800	59,592	3.11	7.28
48	25,600	52,200	3.18	
49	44,800	58,976	3.48	2.12
50	30,400	69,963	2.84	2.23
51	4,800	23,748	3.69	
52	17,600	52,721	3.16	11.95
53	19,200	18,794	2.33	
54	36,800	24,057	2.86	5.01
55	91,200	35,222	2.82	
56	6,400	52,674	2.89	64.93
57	56,000	80,784	3.30	1.76
58	60,800	9,340	2.36	
59	43,200	45,300	2.92	
60	52,800	59,855	2.81	
61	4,800	43,041	3.66	44.27
62	36,800	93,855	2.88	
63	36,800	5,468	2.36	
64	14,400	30,012	3.23	
65	30,400	52,343	4.11	7.53
66	1,600	54,178	4.40	188.58
67	56,000	39,947	3.94	
68	36,800	1,076	6.53	
69	83,200	53,004	4.17	1.53
70	86,400	28,606	4.62	
71	67,200	60,017	3.81	5.05
72	40,000	57,172	3.75	4.31
73	104,000	97,906	3.87	
74	43,200	43,871	3.48	3.03

Table A6 (continued)

District	Cropped Area (hectares)		Population–Cropped-Area Density	
	1960	1980	1960	1980
75	86,400	66,100	3.38	1.48
76	22,400	70,143	4.41	
77	94,400	64,538	3.57	1.60
78	1,440,008	55,279	4.70	0.05
79	51,200	73,346	3.32	3.97
80	52,800	10,381	2.91	
81	49,600	60,151	3.37	3.29
82	60,800	59,657	4.08	
83	49,600	54,732	5.20	2.77
84	81,600	69,963	4.56	3.06
85	8,000	64,599	5.15	27.33
86	57,600	63,905	3.64	3.24
87	75,200	9,680	16.50	1.19
88	41,600	42,263	14.35	9.14
89	60,800	61,599	4.99	3.56
90	94,400	84,736	4.71	1.73
91	104,000	44,871	3.36	1.27
92	48,000	59,179	2.73	2.57
93	57,600	58,846	3.12	1.81
94	51,200	69,323	5.21	9.54
95	59,200	83,181	5.20	
96	100,800	33,313	2.65	
97	73,600	48,838	2.82	2.23
98	75,200	71,322	2.98	
99	96,000	52,393	3.05	0.93
100	40,000	38,355	9.73	5.26
101	78,400	86,385	4.39	
102	40,000	49,468	2.62	
103	78,400	65,733	3.02	2.48
104	40,000	46,816	2.77	
105	52,800	81,944	4.20	
106	52,800	123,070	4.34	
107	59,200	71,748	1.68	
108	51,200	41,392	2.23	8.26
109	0	65,548	2.35	
110	96,000	70,954	2.51	1.30
111		126,632	3.79	

District	Cropped Area (hectares)		Population–Cropped-Area Density	
	1960	1980	1960	1980
112		91,300	3.47	
113	290,700			2.15
114		112,416	10.34	
115				
116				
117				
118				
119				
120				

Notes

PREFACE

1. Jack Ives and Bruno Messerli, *The Himalayan Dilemma: Reconciling Development and Conservation* (London and New York: Routledge, 1989), 98.

CHAPTER 1: HIMALAYA UNDER SIEGE

1. Among the early standard references for Himalayan geology are Augusto Gansser, *Geology of the Himalaya* (London: Interscience Publisher, 1964); Toni Hagen, *Nepal* (Bern: Kummerly and Frey, 1960); and various reports from the Wadia Institute of Himalayan Geology in Dehra Dun, India. More recent overviews include *Geology and Tectonics of the Himalaya* (special publication no. 26 of the Geological Survey of India, Delhi: Government of India, 1989); and John F. Schroder (ed.), *Himalaya to the Sea: Geology, Geomorphology, and the Quaternary* (London and New York: Routledge, 1993). The latter book focuses on northern Pakistan but has relevancy for the wider region. A summary of geological research appears in David Zurick, "The Himalayas," in F. N. Magill (ed.), *Survey of Earth Sciences* (Pasadena, Calif.: Salem Press, 1990), 1073–78. A well-illustrated, popular account of Himalayan geology is

included in Blanche Olschak, Augusto Gansser, and Emil Buhrer, *Himalayas* (New York: Facts on File Publications, 1987).

2. Erik Eckholm, *Losing Ground: Environmental Stress and World Food Production* (New York: Norton, 1976), 77–78. Eloquently written and logically argued, Eckholm's account was one of the early persuasive descriptions of the impending "ecological crisis" in Nepal. It prompted other powerful documentaries, including the 1982 film by Sandra Nichols, *The Fragile Mountain*. See also A. R. Joshi and S. L. Shrestha, "Environmental Impact of Development," *MAB Bulletin*, no. 9 (1988; Kathmandu: Nepal National Committee for Man and the Biosphere).

3. *Newsweek*, November 9, 1987. Quoted in Jack Ives and Bruno Messerli, *The Himalayan Dilemma: Reconciling Development and Conservation* (London and New York: Routledge, 1989), xix. (See note 6.)

4. B. D. Dhawan, "Coping with Floods in Himalayan Rivers," *Economic and Political Weekly* 28, no. 18 (1993):849–53, 849. The article also makes the important point that linking floods to deforestation is a tenuous proposition and is in conflict with current scientific evidence that shows forest loss not to be such an important factor in contributing to floods in the plains. Instead, the seismicity and high precipitation rates of the Himalaya cause floods, regardless of tree cover. For contrasting viewpoints, see R. S. Junor, "Impact of Erosion on Human Activities, Bagmati Catchment, Nepal," *Water Resources Journal*, March 1982; C. L. Shrestha, "River Sedimentation and Remedial Measures in Nepal," *Water Resources Journal*, September 1986; B. Biswas and C. V. Bhardram, "A Study of Major Rainstorms of the Teesta Basin," *Mausam* 35, no. 2 (1984):187–90.

5. Michael Thompson and Thomas Warburton, "Uncertainty on a Himalayan Scale," *Mountain Research and Development* 5, no. 2 (1985):115–35. In this provocative article, the authors brought attention to the fact that a great deal of inaccuracy and uncertainty characterizes Himalayan environmental literature. They believe that such uncertainty itself may be the region's overriding problem. The state of empirical knowledge does not, in their view, allow an accurate assessment of the kinds of crisis events commonly reported in the literature. Yet these same reports have greatly influenced environmental policy makers in the Himalayan countries. According to the authors, the real problems of the Himalaya, and sound policies to remedy them, are diverted by the false claims of an ecological "supercrisis." See also John Metz, "A Reassessment of the Causes and Severity of Nepal's Environmental Crisis," *World Development* 19, no. 7 (1991):805–20.

6. Ives and Messerli, *The Himalayan Dilemma*, 9. The various accounts of ecological collapse and the upland-forest–lowland-flooding linkage that appeared in the 1970s and 1980s literature were combined by Ives and Messerli into a grand Himalayan environmental degradation theory. The development of that theory was supported by numerous articles appearing in the journal *Mountain Research and Development*, especially in vol. 7, no. 3: "Himalayas-Ganges Problem: Proceedings of a Conference" (held in New Paltz, New York, at Mohonk Mountain House, April 1986). The Mohonk Conference, as it has come to be known, focused on several years of United Nations fieldwork in the Himalaya, mainly in Nepal, and debated the issues that gave rise to the Himalaya "supercrisis" scenario.

The outcome of that conference was a fresh look at environmental problems in the Himalaya, leading to the eventual publication of *The Himalayan Dilemma*. That book became a major turning point in Himalaya studies as it constructed and then "debunked" the theory of Himalayan environmental degradation. It showed that while the issue of Indian floods has no legitimate claim on the actions of Himalayan farmers, serious environmental and human problems did exist in the Himalaya. A more recent report by the International Center for Integrated Mountain Development (ICIMOD) in Kathmandu acknowledges the arguments made in the book but also admits that serious problems of nature and society exist in the mountains, many of which are tied to the "supercrisis" scenario:

> Environments and livelihoods are threatened by an increasing imbalance between population and available productive land. In many places the carrying capacity of the land has now been exceeded, leading to an ever-increasing demand for new agricultural and forest land and land-based products. Consequently, the forested upper slopes of these young mountains are being cleared for cultivation, grazing fodder, fuelwood, and timber. Removal of vegetation on steep slopes in conjunction with intense monsoon rainfall is triggering massive erosion and landslides with resulting soil impoverishment and soil losses and a deteriorating biophysical environment. (*Sustainable Mountain Development: ICIMOD Overview*, 1995 [Kathmandu, Nepal], 1)

7. Kenneth Mason, "The Himalaya as a Barrier to Modern Communication," *The Geographical Journal of the Royal Geographical Society* 8, no. 8 (1936):1–16; Nigel Allan, "Real and False Geographies of the Himalaya," *Himal* 5, no. 5 (1992): 20–22.

8. We utilize the insights of political ecology to study the linkage between local environmental management systems, environmental quality, and the political economy of regional development in the Himalaya. The political-ecology perspective integrates ecological, social, and political-economy modes of inquiry. It focuses on interrelationships between local environmental management and the larger political economy, especially on how the distribution of power influences resource decisions. As a hermeneutic device, political ecology is less a cohesive theory and more an amalgam of compatible points of view. It has been employed by geographers in diverse world settings to examine human pressures on production, community resource allocation, poverty and resource depletion, native peoples and parks, and other related issues. Piers Blaike and Harold Brookfield articulate the political-ecology perspective in their 1987 book, *Land Degradation and Society* (London: Methuen), and apply it to problems of environment and declining agricultural productivity in Nepal. Elsewhere in the Himalaya literature, the interests of political ecology are widely shared, but the explicit use of political ecology as a mode of inquiry is not generally made. See for example, Barry Bishop, *Karnali Zone under Stress* (Chicago: University of Chicago Press, 1990); Barbara Brower, *Sherpa of Khumbu* (Delhi: Oxford University Press, 1991); and Stanley Stevens, *Claiming the High Ground* (Berkeley: University of California Press, 1993).

9. Derek Denniston, "Saving the Himalaya," *Worldwatch* 6, no. 6 (1993):10–21, 11–12. See also Derek Denniston, *High Priorities: Conserving Mountain Ecosystems and Cultures* (Washington, D.C.: Worldwatch Institute paper no. 123, 1995).

10. Jayanta Bandyopadhyay, *Status of the World's Mountains: A Global Status Report* (London: Zed Books, 1992).

11. Ramachandra Guha, *The Unquiet Woods: Ecological Change and Peasant Resistance in the Himalaya* (Berkeley: University of California Press, 1990).

CHAPTER 2: MOUNTAIN CONTOURS

1. An abbreviated regional schematic of the Himalaya based upon geoecological factors is introduced in Jack Ives and Bruno Messerli, *The Himalayan Dilemma* (London and New York: Routledge, 1989), 36–42.

2. See, for example, J. F. Dobremez, *Le Nepal: Ecologie et biogeographie* (Paris: Centre National de la Recherche Scientifique, 1976); J. Kawakita, "Vegetation," in H. Kihara (ed.), *Land and Crops of Nepal Himalaya*, vol. 2 (Kyoto: Kyoto University, 1956), 1–65; U. Schweinfurth, *Die horizontale und vertikale Verbreitung der Vegetation in Himalaya* (Bonn: Fred Dummlers Verlag, 1957). A classic study of vegetation in southern Asia, including the Himalaya region, is provided in A. T. A. Learmonth, *The Vegetation of the Indian Sub-Continent*, Occasional Paper no. 1, Department of Geography, Australia National University, Canberra, 1964.

3. S. C. Joshi, *Nepal Himalaya: Geoecological Perspectives* (Nainital: Himalayan Research Group, 1986); Carl Troll, "Comparative Geography of the High Mountains of the World in the View of Landscape Ecology," in Nigel Allan, Gregory Knapp, and C. Stadel (eds.), *Human Impact on Mountains* (Totowa, N.J.: Rowman Littlefield, 1988). References to Harald Uhlig's work in the Himalaya also appear in *Human Impact on Mountains*. The 1995 conference Karakorum-Hindu Kush-Himalaya: Dynamics of Change, held in Islamabad under the auspices of Tübingen University in Germany and the Pakistan Government, reported on five years of geoecological research in northern Pakistan primarily by German researchers.

4. Nigel Allan, "Accessibility and Altitudinal Zonation Models of Mountains," *Mountain Research and Development* 6, no. 3 (1986):185–94; Melvyn Goldstein and Donald Messerschmidt, "The Significance of Latitudinality in Himalayan Mountain Ecosystems," *Human Ecology* 8, no. 2 (1980):117–34; David Guillet, "Toward a Cultural Ecology of Mountains: The Central Andes and the Himalayas Compared," *Current Anthropology* 24, no. 5 (1983):561–74.

CHAPTER 3: THE SHAPE OF LAND AND LIFE

1. The literature on sustainable development is vast, but see, for examples, Susanna Hecht, "Environment, Development, and Politics," *World Development* 13(1985):663–84; Michael Redclift, *Sustainable Development* (London: Methuen, 1987); and Michael Watts, "Sustainable Development and Struggles over Nature," in F. Buttel and L. Thrupp (eds.), *The Political Economy of Sustainable Development* (Ithaca, N.Y.: Cornell University Press, 1992). Examples of the recent application of sustainable-development principles in the Himalaya are found in *Bhutan: Towards Sustainable Development in a Unique Environment*, National Report for the United Nations Conference on Environment and Development (UNCED; Thimphu, n.d.); International Center for Integrated Mountain Development, *Mountain Environment and Development* (symposium proceedings; Kathmandu: ICIMOD, 1994); P. P. Karan, Hiroshi Ishii et al., *Nepal: Development and Change in a Landlocked Himalayan Kingdom* (Tokyo: Tokyo University of Foreign Studies, Monumenta Serindica no. 25, 1994).

2. International Center for Integrated Mountain Development, *Our Mountains—*

the Hindu Kush—Himalayas: A Decade of Efforts toward Integrated Mountain Development, 1983–1993 (Kathmandu: ICIMOD, 1993), 1. The Center (ICIMOD) was established in 1983 in Kathmandu, Nepal, as a research and training facility for the entire Himalaya region. Its country members include Afghanistan, Bangladesh, Bhutan, China, India, Myanmar, Nepal, and Pakistan. ICIMOD advocates an integrated approach to resolve the region's diverse problems and organizes its activities around the problem areas of mountain farming systems, mountain enterprises and infrastructure, and mountain natural resources. It provides three main services: information, training, and project administration. The ICIMOD offices are located at 4/80 Jawalakhel, GPO Box 3226, Kathmandu, Nepal.

3. Royal Government of Bhutan, Seventh Five-Year Plan (1992–1997), vol. 1, Main Plan Document (Thimphu, 1991): Planning Commission, Royal Government of Bhutan), xviii. An example of Bhutan's commitment to sustainable development is the Bhutan Trust Fund for Environmental Conservation, created in 1991 in partnership with government and donors to finance environmental protection programs in the country.

4. Modern social theory proposes that the analysis of human society should consider the agencies of local people. By analogue, the explication of landscape change requires the study of particular places. Our study of the Himalaya considers both these elements as part of a postmodern regional geography, wherein nature and society are co-dependents. The relations that exist between them reflect continual cultural appraisals of both the natural environment and economic capacities. The decision making, however, shifts with the scale of interaction—from the village to the global economy. For a general discussion of the importance of geographic scale, see William Meyer, Derek Gregory et al., "The Local-Global Continuum," in Ronald Abler, Melvin Marcus, and Judy Olson (eds.), Geography's Inner Worlds (New Brunswick, N.J.: Rutgers University Press, 1992).

5. Thomas Fricke, "Introduction: Human Ecology in the Himalaya," Human Ecology 17, no. 2 (1989):131–45 (special volume of papers by geographers and anthropologists on issues of cultural ecology in Nepal).

6. David Guillet, "Toward a Cultural Ecology of Mountains: The Central Andes and the Himalayas Compared," Current Anthropology 24, no. 5 (1983):561–74; Benjamin Orlove and David Guillet, "Theoretical and Methodological Considerations in the Study of Mountain Peoples: Reflections on the Idea of Subsistence Type and the Role of History in Human Ecology," Mountain Research and Development 5, no. 1 (1985):3–18.

7. The regional level of inquiry is advocated in the political-ecology approach articulated in Piers Blaikie and Harold Brookfield, Land Degradation and Society (London: Methuen, 1987).

8. Advocates of the "new regional geography" dismiss the particularistic and descriptive studies of an earlier generation of regional geographers as being devoid of social theory and explanatory power, as being static and isolated, and therefore as having little theoretical promise. Instead, they propose a reconstructed, more radical approach to regional study that describes not only the bounded context of daily life (analytical regions or scales of analysis) but also the mix of global and local processes that gives shape to it. For an assessment of the main arguments of the new regional geography, see: Hans Holmer, "What's New and What's Regional in the 'New

Regional Geography?'" *Geografiska annaler* 77, no. 1 (1995):47–62; Mary Beth Pudup, "Arguments within Regional Geography," *Progress in Human Geography* 12, no. 3 (1988):369–90; A. Sayer, "The 'New' Regional Geography and Problems of Narrative," *Environment and Planning D: Society and Space* 7, no. 3 (1989):253–76; and a series of articles by Nigel Thrift in *Progress in Human Geography* (1990, 1991, and 1993).

9. See, for example, Blanch Olschak, Augusto Gansser, and Emil Buhrer, *Himalayas* (New York: Facts on File Publications, 1987).

10. Derek Denniston, *High Priorities: Conserving Mountain Ecosystems and Cultures* (Washington, D.C.: Worldwatch Institute paper no. 123, 1995).

11. Quoted in Derek Denniston, "Saving the Himalaya," *Worldwatch* 6, no. 6 (1993):10–21, 12.

12. Pradyumna Karan, "Geographic Regions of the Himalayas," *Bulletin of Tibetology* 3, no. 2 (1966):1–25.

13. For a theoretical overview of Himalayan subsistence types see Benjamin Orlove and David Guillet, "Theoretical and Methodological Considerations in the Study of Mountain Peoples: Reflections on the Idea of Subsistence Type and the Role of History in Human Ecology," *Mountain Research and Development* 5, no. 1 (1985): 3–18. Detailed studies of subsistence types in the Himalaya include Gerald Berreman, "Ecology, Demography and Domestic Strategies in the Western Himalaya," *Journal of Anthropological Research* 34, no. 2 (1978):326–68; Barry Bishop, *Karnali Zone under Stress* (Chicago: University of Chicago Press, 1990); Barbara Brower, *Sherpa of Khumbu* (New Delhi: Oxford University Press, 1991); Tom Fricke, *Himalayan Households* (New York: Columbia University Press, 1994); John Metz, "A Framework for Classifying Subsistence Production Types in Nepal," *Human Ecology* 17, no. 2 (1989):131–45; and Stan Stevens, *Claiming the High Ground* (Berkeley: University of California Press, 1993).

14. David Zurick, "The Road to Shangri La Is Paved: Spatial Development and Rural Transformation in Nepal," *South Asia Bulletin* 13, nos. 1 and 2 (1993):35–44.

CHAPTER 4: MYTH AND PREHISTORICAL TERRITORY

1. S. S. Charak, *History and Culture of Himalayan States*, vol. 2 (New Delhi: Light and Life Publishers, 1979). See also A. B. Mukerji, "Rural Settlements of the Chandigarh Siwalik Hills (India): A Morphogenetic Analysis," *Geografiska Annaler, Series B* 58, no. 2 (1976):95–115.

2. The anthropologist Gerald Berreman suggests that the strong component of *khasa* groups in the contemporary ethnic profile of Himalayan communities in the Kumaun region indicates a more widespread earlier distribution. See Gerald Berreman, *Hindus of the Himalaya* (Berkeley: University of California Press, 1972). Khasa migration into the central Himalaya and the subsequent domination of the aboriginal people is rebutted, however, by some cultural historians. They view the khasa migration theory as being primarily an elitist explanation of the origins of the central Himalayan society and one that does poor service to aboriginal history. See, for example, Maheshwar P. Joshi, "Culture Constructed by Intellectualism and the Intellectualism of Culture: The Case of the Central Himalaya," a paper presented at the International Symposium on Karakoram-Hindu Kush-Himalaya: Dynamics of Change, Sept. 29–Oct. 2, 1995, in Islamabad, Pakistan.

3. Archaeological work by Tibetologist Giuseppi Tucci provides inscriptural evidence of the range of khasa dominance in western Nepal and adjoining areas of Garhwal in India. Giuseppi Tucci, *Nepal: The Discovery of the Malla* (New York: Dutton Publishers, 1962).

4. Ludwig Stiller, *The Rise of the House of Gorkha* (New Delhi: Manjusri Publishing House, 1973).

5. Richard English, "Himalayan State Formation and the Impact of British Rule in the Nineteenth Century," *Mountain Research and Development* 5, no. 1 (1985):61–78; A. C. Talukdar, "Political Evolution of Arunachal Pradesh," *Proceedings of Northeast India Historical Association* (Pasighat, Arunachal Pradesh, 1987), 131–54.

6. S. C. Bajpai, *Lahaul-Spiti* (New Delhi: Indus Publishing Company, 1987).

7. John H. Crook, "The History of Zangskar," in John Crook and Henry Osmaston (eds.), *Himalayan Buddhist Villages* (New Delhi: Motilal Banarsidass Publishers, 1994). This publication is a major summation of several years of collaborative work undertaken by researchers associated with the Zanzkar Project. It is perhaps one of the most comprehensive interdisciplinary studies ever published about the western Himalaya region. See also Harjit Singh, "Ladakh: Socio-economic Changes and Current Disturbances," a paper presented at a seminar titled "Disturbed Ladakh: Its Socio-economic and Geographical Dimensions," held at the Center for the Study of Regional Development, Jawaharlal Nehru University, October 27, 1989.

8. Christophe von Furer-Haimendorf, *Himalayan Traders* (on the Arun Valley in eastern Nepal; New York: St. Martins Press, 1975); Tom Fricke, *Himalayan Households* (on the upper Ankhu Khola in central Nepal; New York: Columbia University Press, 1994); Barbara Brower, *Sherpa of Khumbu* (on the Khumbu Valley in northeast Nepal; New Delhi: Oxford University Press, 1991); Stan Stevens, *Claiming the High Ground* (also on the Khumbu Valley; Berkeley: University of California Press, 1993).

CHAPTER 5: THE MEDIEVAL ERA OF MOUNTAIN PRINCES

1. David Zurick, 1989, "Historical Links between Settlement, Ecology, and Politics in the Mountains of West Nepal," *Human Ecology* 17, no. 2 (1989):229–55; Mahesh C. Regmi, *Thatched Huts and Stucco Palaces: Peasants and Landlords in Nineteenth-Century Nepal* (New Delhi: Vikas Publishing House, 1977); Ludwig Stiller, *The Rise of the House of Gorkha* (New Delhi: Manjusri Publishing House, 1973).

2. Pradyumna P. Karan, *Bhutan: A Physical and Cultural Geography* (Lexington: University of Kentucky Press, 1967); Leo Rose, *The Politics of Bhutan* (Ithaca: Cornell University Press, 1977).

3. Richard English, "Himalayan State Formation and the Impact of British Rule in the Nineteenth Century," *Mountain Research and Development* 5, no. 1 (1985):61–78.

4. Mahesh C. Regmi, *A Study of Nepalese Economic History, 1776–1847* (New Delhi: Munjushri Publishing House, 1971).

5. Richard P. Tucker, "The British Colonial System and the Forests of the Western Himalaya, 1815–1914," in Richard P. Tucker and J. F. Richards (eds.), *Global Deforestation and the Nineteenth-Century World Economy* (Durham, N.C.: Duke University Press, 1983), 146–66. For a detailed discussion of the forest history of a community in central Nepal, see T. B. S. Mahat, D. M. Griffin, and K. R. Shepherd, "Human Impact on Some Forests of the Middle Hills of Nepal," I: "Forestry in the context of the

traditional resources of the state," and II: "Some major human impacts before 1950 on the forests of Sindhu Palchok and Kabhre Palanchok," in, respectively, *Mountain Research and Development* 6, no. 3 (1986):223–32, and *Mountain Research and Development* 6, no. 4 (1986):325–34.

6. Lionel Caplan, *Administration and Politics in a Nepalese Town* (London: Oxford University Press, 1975).

7. Leo Rose and John Scholz, *Nepal: Profile of a Himalayan Kingdom* (Boulder: Westview Press, 1980); Lionel Caplan, *Land and Social Change in East Nepal* (Berkeley and Los Angeles: University of California Press, 1970).

CHAPTER 6: THE COLONIAL GREAT GAME

1. Mary Des Chene persuasively argues that by allowing the recruitment of Nepalese hill tribes for the British colonial army, the Rana ruling family bolstered its own political position in Nepal. This is in keeping with other forms of alliance between the Rana oligarchy and the British raj. See Mary Des Chene, "Soldiers, Sovereignty and Silences: Gorkhas as Diplomatic Currency," *South Asia Bulletin* 13, no. 1/2 (1993):67–80. Also, Lionel Caplan, *Warrior Gentlemen: Gurkhas in the Western Imagination* (Oxford: Berghahn Books, 1995).

2. Alexander MacKenzie, *The Northeast Frontier of India* (New Delhi: Mittal Publications, 1979). (First published in 1884 as "History of the Relations of the Government with the Hill Tribes of the Northeast Frontier of Bengal.")

3. *Proceedings of Northeast India Historical Association* (Pasighat and Shillong, Arunachal Pradesh, 1987).

4. George L. Harris et al., *Area Handbook for Nepal, Bhutan, and Sikkim* (Washington, D.C.: U.S. Government Printing Office, 1973); Pradyumna Karan, *Bhutan: Development amid Environmental and Cultural Preservation* (Tokyo: Tokyo University of Foreign Studies, Institute for the Study of Languages and Cultures of Asia and Africa, Monumenta Serindica no. 17, 1987); Ram Rahul, *The Himalaya as a Frontier* (New Delhi: Vikas Publishing House, 1978); N. R. Roy (ed.), *Himalaya Frontier in Historical Perspective* (Calcutta: Institute of Historical Studies, 1986). Pradyumnan Karan, *Sikkim Himalaya: Development in a Mountain Environment* (Tokyo: Tokyo University of Foreign Studies, Institute for the Study of Languages and Cultures of Asia and Africa, Monumenta Serindica no. 13, 1984).

5. For an excellent and highly readable account of exploratory travel by British commercial agents and spies in the northwest Himalaya and Karakoram region, see John Keay, *Where Men and Mountains Meet* (Karachi: Oxford University Press, 1993). The book frames the early travel adventures in the context of the Great Game played out by the British, the Russians, and the Chinese.

6. Quoted in Ian Cameron, *Mountains of the Gods: The Himalaya and the Mountains of Central Asia* (London: Century Hutchinson, 1984), 85.

7. Quoted in Cameron, 109–10.

8. Matthew H. Edney, "The Patronage of Science and the Creation of Imperial Space: The British Mapping of India, 1799–1843," *Cartographica* 30, no. 1 (1993): 61–67.

9. Quoted in Ranesh C. Kalita, "British Exploitation in Assam: The Opium Policy and Revenue, 1850–94," in *Proceedings of Northeast India History Association, Twelfth Session* (Shillong, Arunachal Pradesh, 1991), 344. The Assam trade routes are de-

scribed in H. K. Barujari, *Assam in the Days of the Company (1826–1858)* (Gauhati, Assam: Spectrum Publications, 1980); Nisar Ahmad, "Economy of Ancient Assam: A Review in Historical Perspective," in Milton S. Sangma (ed.), *Essays on Northeast India* (New Delhi: Indus Publishing Company, 1994), 93–107; Chittabrata Palit, "British Economic Penetration into Northeastern Hills: Overland Trade and Allied Questions, 1800–1850," in N. R. Roy (ed.), *Himalaya Frontier in Historical Perspective* (Calcutta: Institute of Historical Studies, 1986), 102–12.

10. Mahesh C. Regmi, *The State and Economic Surplus: Production, Trade, and Resource Mobilization in Early 19th-Century Nepal* (Varanasi: Nath Publishing House, 1984).

11. Brian Hodgeson, *Papers Relative to the Colonization, Commerce, Physical Geography, Science, and Culture of the Himalaya Mountains and Nepal: Selections from the Records of the Government of Bengal, no. 26* (Calcutta, 1857); see also J. Pemble, *The Invasion of Nepal: John Company at War* (Oxford: Clarendon Press, 1971).

12. For excellent overviews of the historical links between global colonialism, commerce, and impacts on local economy, see, for example, John Isbister, *Promises Not Kept: Betrayal of Social Change in the Third World* (West Hartford, Conn.: Kumerian Publishers, 1993), esp. chap. 4; and Jim Blaut, *The Colonizers' Model of the World* (New York and London: Guilford Press, 1993).

13. A comprehensive study of geopolitics and travel routes during the colonial and postcolonial periods in the western Himalaya is found in Ispahani Mahnaz, *Roads and Rivals: The Political Uses of Access in the Borderlands of Asia* (Cornell: Cornell University Press, 1989). See also Nigel Allan, "Real and False Geographies of the Himalaya," *Himal* 5, no. 5 (1992):20–22; Kenneth Mason, "The Himalaya as a Barrier to Modern Communications—South Asia Lecture," *The Geographical Journal* 87, no. 1 (1936):1–16.

14. Estimates for Dehra Dun are provided in J. F. Richards, "Environmental Change in Dehra Dun Valley, India: 1860–1980," *Mountain Research and Development* 7, no. 3 (1987):299–304. The Kumaun estimates are reported in Richard P. Tucker, "The British Empire and India's Forest Resources: The Timberlands of Assam and Kumaun, 1914-1950," in John F. Richards and Richard P. Tucker (eds.), *World Deforestation in the Twentieth Century* (Durham and London: Duke University Press, 1988).

15. Haripriya Rangan, "Contested Boundaries: State Policies, Forest Classifications, and Deforestation in the Garhwal Himalayas," *Antipode* 27, no. 4 (1995):343–62; Sanjeev Prakash, "Social Institutions and Common Property Resources in the Mountains," *Mountain Research and Development* 18, no. 1 (1998):1–3.

16. Percival Griffiths, *The History of the Indian Tea Industry* (London: Weidenfeld and Nicholson, 1967).

CHAPTER 7: A DIVIDED GEOGRAPHY

1. The International Symposium on Karakorum-Hindu Kush-Himalaya: Dynamics of Change—held Sept. 29–Oct. 2, 1995, in Islamabad, Pakistan—debated the impact of the Karakoram Highway on the Indus mountains in northern Pakistan. Numerous research reports presented at the conference documented changes in land use, economy, society, and the environment that are attributed to the completion of the highway.

2. P. P. Karan, Hiroshi Ishii et al., *Nepal: Development and Change in a Landlocked*

Himalayan Kingdom (Tokyo: Tokyo University of Foreign Studies, Monumenta Serindica no. 25, 1994).

3. Christopher Flavin, "Sustainable Development in the Kingdom of Bhutan," discussion paper for the national workshop on sustainable development and the environment (Paro: Royal Government of Bhutan, UNDP, and DANIDA, 1990); Planning Commission, Royal Government of Bhutan, *Seventh Five-Year Plan (1992–1997)*, vol. 1, *Main Plan Document* (Thimphu, 1991); Department of Forestry, Royal Government of Bhutan, *Master Plan for Forestry Development in Bhutan: Wood Energy Sectoral Analysis* (Bangkok: FAO document, 1991).

4. Prem Thadhani, *Chronicles of the Doon Valley: An Environmental Exposé* (New Delhi: Indus Publishing Company, 1993); J. Bandyopadhyay, *Natural Resource Management in the Mountain Environment: Experiences from the Doon Valley, India* (Kathmandu: ICIMOD, 1989).

5. Gerald Berreman, "Uttarkhand and Chipko: Regionalism and Environmentalism in the Central Himalaya," in Manis K. Raha (ed.), *The Himalayan Heritage* (Delhi: Gian Publishing House, 1987); Vijay Paranjpye, *Evaluation of the Tehri Dam* (New Delhi: Indian National Trust for Art and Cultural Heritage, 1988).

6. Ramchandra Guha, *The Unquiet Woods: Ecological Change and Peasant Resistance in the Himalaya* (Berkeley: University of California Press, 1990); Manishal Aryal, "Axing Chipko," *Himal* 7, no. 1 (1994):8–23; Pradyumna Karan, "Environmental Movements in India," *Geographical Review* 84, no. 1 (1994):32–41; Paul Harris, "Tribal Gatherings on the Northeast Frontier," *Geographical Magazine*, October 1997, 53–58.

CHAPTER 8: A CHAIN OF EXPLANATION

1. Daily life in the village of Phalabang, Nepal, is described in David Zurick, "A Question of Balance," *Sierra* 72, no. 4 (1987):46–50. The cultural ecology of Phalabang is studied in David Zurick, "Traditional Knowledge and Conservation as a Basis for Development in Western Nepal," *Mountain Research and Development* 10, no. 1 (1990):23–33.

2. A wonderful evocation of the duality between the nearness of cultural hearth and community and the wider spheres of society and space is found in the writings of geographer Yi Fu Tuan. See especially *Cosmos and Hearth: A Cosmopolite Viewpoint* (Minneapolis: University of Minnesota Press, 1996).

3. Jack Ives and Bruno Messerli, *The Himalayan Dilemma* (London and New York: Routledge, 1989).

4. Michael Thompson and Thomas Warburton, 1985, "Uncertainty on a Himalayan Scale," *Mountain Research and Development* 5, no. 2 (1985):115–32.

5. The regional heterogeneity in population pressure and environmental degradation is important, not to discount the problems of environmental carrying capacity but to note its highly localized characteristics. See, for example, P. P. Karan and Shigeru Iijima, "Environmental Stress in the Himalaya," *Geographical Review* 75, no. 1 (1985):71–92.

6. Ives and Messerli, *The Himalayan Dilemma*, 3–4.

7. The materials used to compile the numbers include the Indian Census and the Bureau of Statistics, Government of Nepal. Some of the early census figures, especially for Nepal, are provisional at best, and the early population figures throughout

should be considered only as estimates. The breakdown of population by subregion is as follows:

	1890	1950	1990
Western Himalaya	11,361,692	15,403,462	29,092,002
Nepal	5,638,749	8,473,478	18,491,097
Darjeeling	223,314	445,260	1,024,269
Sikkim	30,458	137,725	316,385
Bhutan and E. Himalaya	no data	336,558	1,662,157
Total	17,254,213	24,796,483	50,585,910

8. Eric Eckholm, *Losing Ground* (New York: W. W. Norton and Company, 1976).

9. See, for example, Lawrence Hamilton, "What Are the Impacts of Himalayan Deforestation on the Ganges-Brahmaputra Lowlands Delta? Assumptions and Facts," *Mountain Research and Development* 7, no. 3 (1987):256–63, 262; Brian Carson, *Erosion and Sedimentation Processes in the Nepalese Himalaya* (Kathmandu: ICI-MOD Occasional Paper no. 1, 1985), 36; Bruno Messerli and Thomas Hofer, "Assessing the Impact of Anthropogenic Land Use Change in the Himalaya," in Graham Chapman and Michael Thompson (eds.), *The Quest for Sustainable Development in the Ganges Valley* (London: Mansell Publishing Co., 1995), 64–89, 87. A great deal of recent literature has been produced on the tenuous linkage between upland landscape processes and lowland surface hydrology. See, for example, Graham Chapman and Michael Thompson, *Water and the Quest for Sustainable Development in the Ganges Valley* (London: Mansell Publishing Co., 1995); and for a study of the appropriate scale of biophysical models, see Douglas Gamble and Vernon Meentmeyer, "The Role of Scale in Research on the Himalaya-Ganges-Brahmaputra Interaction," *Mountain Research and Development* 16, no. 2 (1996):149–55.

10. B. D. Dhawan, "Coping with Floods in Himalayan Rivers," *Economic and Political Weekly* 28, no. 18 (1993):849–53. The issue brings with it a host of related matters, including the need for human resettlement, the role of dams in both producing and mitigating floods, and the value of integrated watershed management programs.

11. Barry Bishop, *Karnali Zone under Stress* (Chicago: University of Chicago, 1990).

12. Deepak Bajracharya, "Fuel, Food, or Forest? Dilemmas in a Nepali Village," *World Development* 11, no. 12 (1983):1057–74. The contention over fuelwood use rates was a highly visible one in the scientific literature and in the government reports during the 1970s and 1980s. The "67" factor was reported in *The Himalaya Dilemma* and is based on a consultancy report by Deanna Donavan, "Fuelwood: How Much Do We Need?" *Newsletter DGD 14*, Institute for Current Affairs, Hanover, New Hampshire, 1981. See as well R. L. Shrestha, "Socioeconomic Factors Leading to Deforestation in Nepal" (Kathmandu: Research and planning paper series no. 2, HMG-USAID-GTZ-IDRC-Winrock Project, 1986).

13. John Metz, 1991, "A Reassessment of the Causes and Severity of Nepal's Environmental Crisis," *World Development* 19, no. 7 (1991):805–20.

1. For a compelling discussion of natural hazards in mountain areas, see Kenneth Hewitt, "Hazards and Disasters in Mountain Environments: Problems in the Geography of Risk," in Nigel J. R. Allan (ed.), *Mountains at Risk* (New Delhi: Manohar Publishers, 1995), 98–128. See also A. B. Mukerji, "Geomorphic Damages in the Western Himalaya: Regional Patterns and Some Human Dimensions Revealed by Newspaper Reports," in *Erdwissenschaftliche Forschung, Bd. XVIII* (1984), 127–47.

2. Vinod Gaur, *Earthquake Hazard and Large Dams in the Himalaya* (New Delhi: Indian National Trust for Art and Cultural Heritage [INTACH], 1993); R. S. Mithal, B. C. Joshi, and K. Gohain, "Environmental Impacts of the Ramganga Dam Project," *The National Geographical Journal of India* 30, no. 3 (1984):81–91.

3. M. B. Fort and P. Freytet, "The Quarternary Sedimentary Evolution of the Intra-Montane Basin of Pokhara in Relation to the Himalaya Midlands and then Hinterlands (West Central Nepal)," in A. K. Sinha (ed.), *Contemporary Geoscientific Researches in the Himalaya, Vol. 2* (Dehra Dun, 1982), 91–96—reported in Jack Ives and Bruno Messerli, *The Himalayan Dilemma: Reconciling Development and Conservation* (London and New York: Routledge, 1989).

4. *Master Plan for the Forestry Service Project (MPFS). Main Report: Forest Development Plan for the Supply of Main Forest Products, Forest Resources Information Status, and Development Plan for the Conservation of Ecosystems and Genetic Resources* (Kathmandu: HMGN/FINNIDA/ADB, 1988).

5. Alton Byers, "A Geomorphic Study of Man-Induced Soil Erosion in the Sagarmatha (Mount Everest) National Park, Khumbu, Nepal," *Mountain Research and Development* 7, no. 3 (1987):209–16.

6. The "wildness" of nature, including the potential for natural hazards, is incorporated into native peoples' overall conception of the landscape. D. Vuichard and M. Zimmermann, "The Catastrophic Drainage of a Moraine-Dammed Lake, Khumbu Himal, Nepal: Cause and Consequences," *Mountain Research and Development* 7, no. 2 (1987):91–110.

7. B. B. Deoja, "Planning and Management of Mountain Roads," *Proceedings of the International Symposium on Mountain Environment and Development* (Kathmandu: ICIMOD, 1994), 1–44.

8. See, for example, Nigel Allan, "Human Aspects of Mountain Environmental Change," in his *Mountains at Risk*, 3–26. Reference to the watershed studies are provided in Bruno Messerli and Thomas Hofer, 1995, "Assessing the Impacts of Anthropogenic Land Use Change in the Himalaya," in Graham Chapman and Michael Thompson (eds.), *Water and the Quest for Sustainable Development in the Ganges Valley* (London: Mansell Publishing Co., 1995), 64–89.

9. Martin Price and John Haslett, "Climate Change and Mountain Ecosystems," in Allan, *Mountains at Risk*, 73–97; P. C. Adhikary, "The Monsoon Circulation in Nepalese Himalayas: Evolution and Environmental Thrust," *Bulletin, Department of Geology, Tribhuvan University* 3, no. 1 (1993):1–10.

10. Norman Myers, "Threatened Biotas: 'Hot Spots' in Tropical Forests," *The Environmentalist* 8, no. 3 (1988):187–208.

11. Mahesh Banskota and Archana Karki, *Sustainable Development of Fragile Moun-

tain Areas of Asia: Regional Conference Report, December 13–15, 1994 (Kathmandu: ICI-MOD, 1994).

12. For a more detailed discussion of mountain fragility as it relates to agricultural production in the Himalaya, see N. S. Jodha and S. Shrestha, "Sustainable and More Productive Mountain Agriculture: Problems and Prospects," in *Proceedings of the Tenth Anniversary Symposium of the International Center for Integrated Mountain Development, Dec. 1–2, 1993* (Kathmandu, 1994), 1–54.

13. For an in-depth discussion of these matters, see N. S. Jodha, "The Nepal Middle Mountains," in J. Kasperson, R. Kasperson, and B. L. Turner II (eds.), *Regions at Risk* (Tokyo: United Nations University Press, 1995), 140–85.

CHAPTER 10: THE WEIGHT OF LIFE

1. Nigel Allan, *Mountains at Risk* (New Delhi: Manohar Publishers, 1995).

2. More specifically, Nepal has provided the model for the ecocrisis scenario reported for the entire Himalaya.

3. Allan, *Mountains at Risk*.

4. The efforts to link resource conditions with socioeconomic trends have come to employ sophisticated geographical analyses, as shown by the work of the spatial analysis laboratory at the International Center for Integrated Mountain Development in Kathmandu, Nepal. See also *GIS Database of Key Indicators of Sustainable Mountain Development in Nepal*, Mountain Environment and Natural Resources Information Services (MENRIS) (Kathmandu: ICIMOD/UNEP, 1996).

5. Based on 1980 and 1990 district-level figures reported by the governments of India, Pakistan, and Nepal. See also N. S. Jodha and S. Shrestha, "Sustainable and More Productive Mountain Agriculture: Problems and Prospects," in *Proceedings of the Tenth Anniversary Symposium of the International Center for Integrated Mountain Development, Dec. 1–2, 1993* (Kathmandu: ICIMOD, 1994), 1–54.

6. For a detailed analysis of land use change and their impacts on agriculture in Phalabang village and the surrounding regions of the Rapti zone in western Nepal, see David Zurick, "Resource Needs and Land Stress in Rapti Zone, Nepal," *The Professional Geographer* 40, no. 4 (1988):428–44.

7. N. S. Jodha, *A Framework for Integrated Mountain Development* (Kathmandu: ICIMOD discussion paper series MFS no. 1, 1989).

8. One of the earliest and most comprehensive of such efforts was the Canadian-based Land Resources and Mapping Project (LRMP) in Nepal (1985) by Kenting Earth Sciences, Ottawa. Airphoto interpretation and mapping at the 1:50,000 scale contributed to a country database that includes numerical assessments of precipitation, temperature, soil depth, soil fertility, land cover and use, slope steepness, and land capability.

9. Jodha and Shrestha, "Sustainable and More Productive Mountain Agriculture." See also specific country recommendations that appear in Mahesh Banskota and Archana Karki, *Sustainable Development of Fragile Mountain Areas of Asia* (Kathmandu: ICIMOD, 1995). Also, Jacqueline Ashby and Douglas Pachico, "Agricultural Ecology of the Mid-Hills of Nepal," in B. L. Turner II and S. B. Brush (eds.), *Comparative Farming Systems* (New York and London: Guilford Press, 1987), 195–227.

10. A series of fifty-three maps produced by the Geographic Information Systems program of the MENRIS project at ICIMOD show important spatial trends for

Nepal's mountain districts. The maps include coverage of development indicators and infrastructure change (Kathmandu: ICIMOD/MENRIS, 1996).

11. P. P. Karan, Hiroshi Ishii et al., *Nepal: Development and Change in a Landlocked Himalayan Kingdom* (Tokyo: Institute for the Study of Languages and Cultures of Asia and Africa, Monumenta Serindica no. 25, 1994). For a detailed study of land ownership and migration in the tarai, see Nanda Shrestha, *Landlessness and Migration in Nepal* (Boulder: Westview Press, 1990).

12. Planning Commission, Royal Government of Bhutan, *Seventh Five-Year Plan (1992–1997)*, vol. 1, *Main Plan Document* (Thimphu, 1991).

13. For a discussion of this issue in the context of the arid high valleys of northern Pakistan, see Nigel Allan, "Household Food Supply in Hunza Valley, Pakistan" *Geographical Review* 80, no. 4 (1990): 400–15.

14. N. S. Jodha, *Global Changes and Environmental Risks in Mountain Environments* (Kathmandu: ICIMOD discussion paper series MFS no. 23, 1992).

15. See, for example, Government of Bhutan, *Seventh Five-Year Plan (1992–1997)*; Christopher Flavin, "Sustainable Development in the Kingdom of Bhutan," discussion paper for the national workshop on sustainable development and the environment, Paro, Bhutan, May 4–5, 1990; *Bhutan: Towards Sustainable Development in a Unique Environment*, National Report for the United Nations Conference on Environment and Development (Thimphu, n.d.). All of the above reports contain different figures for population and land areas.

16. See, for example, the recent work by U. Schickoff, "Himalayan Forest Cover Change in Historical Perspective: A Case Study in the Kaghan Valley, Northern Pakistan," *Mountain Research and Development* 15, no. 1 (1995):3–18. Also the influential studies on forest conditions that have been conducted in central Nepal under the auspices of the Nepal-Australia Forestry Project: for example, T. Mahat, D. Griffin, and K. Shepherd, "Human Impact on Some Forests in the Middle Hills of Nepal. II: Some major human impacts before 1950 on the forest of Sindhu Palchok and Kabhre Palanchok," *Mountain Research and Development* 7, no. 3 (1986):325–34.

17. Numerous and often widely conflicting estimations of Nepal's forested areas are available. They include J. Martens, *Forests and Their Destruction in the Himalayas of Nepal* (2% per year loss) (Kathmandu: Nepal Research Center Miscellaneous Papers no. 35, 1983); M. D. Joshi, "Environment of Nepal" (3% per year loss), in T. C. Majpuria (ed.), *Nepal—Nature Paradise* (Bangkok: White Lotus Publishers, 1984); Land Resources and Mapping Project (LRMP), *Land Systems Report* (3.3% per year loss) (HMG Nepal and Government of Canada, Kenting Earth Sciences, Ottawa, 1986); *Master Plan for the Forestry Sector Project (MPFS). Main Report* (Kathmandu, 1988; various estimates of loss); Mahesh Banskota et al., *Economic Policies for Sustainable Development in Nepal* (Kathmandu: ADB/ICIMOD, 1990; various estimates of loss).

18. Very little scientific documentation exists for Bhutan. But see Caroline Sargent, Orlando Sargent, and Roger Parsell, "The Forests of Bhutan: A Vital Resource for the Himalaya?" *Journal of Tropical Ecology* 1, no. 4 (1985):265–68; Caroline Sargent, "The Forests of Bhutan," *Ambio* 14, no. 2 (1985):75–80; Food and Agriculture Organization, *Bhutan: Tropical Forest Resources Assessment Project: Forest Resources of Tropical Asia* (Rome: FAO, 1981); and P. P. Karan and Shigeru Iijima, *Bhutan: Development amid Environmental and Cultural Preservation* (Tokyo: Institute for the Study of Languages and Cultures of Asia and Africa, 1987).

19. Forest Survey of India, *The State of Forest Report, 1989* (Dehra Dun: Ministry of Environment and Forest, Government of India, 1989).

20. The forest statistics summarized here are drawn from numerous sources, including those provided by: Ministry of Agriculture, Pakistan; Ministry of Forests, India; India District Gazetteers; Central Bureau of Statistics, Nepal; LRMP, Kenting Earth Sciences, Ottawa, and HMG-Nepal, Kathmandu; Government Planning Commission, Bhutan. Decennial figures were obtained for 120 districts, compiled into a spatial database and analyzed according to numerous cartographic and Geographical Information Systems procedures.

21. The geographers Jack Ives and Bruno Messerli discuss this point at some length in *The Himalayan Dilemma* (London and New York: Routledge, 1989). They link it to the alarming degree of uncertainty about environmental conditions in the mountains.

22. Classical references to the economy and ecology of Himalayan farming communities, most of which center on Nepal, include S. D. Pant, *The Social Economy of the Himalaya* (London: Allen and Unwin, 1935); John Hitchcock, *The Magars of Banyon Hill* (New York: Holt, Rinehart, and Winston, 1966); Gerald Berreman, *The Hindus of the Himalaya* (Berkeley: University of California Press, 1963); J. F. Dobremez and C. Jest, *Manaslu: Hommes et mileux des vallées du Nepal central* (Paris: Centre National de la Recherche Scientifique, 1976); Alan Macfarlane, *Resources and Population: A Study of the Gurungs of Nepal* (Cambridge: Cambridge University Press, 1976); Donald Messerschmidt, *The Gurungs of Nepal: Conflict and Change in a Village Society* (Warminster: Aris and Phillips, 1976); Mark Poffenberger, *Patterns of Change in the Nepal Himalaya* (Delhi: Macmillan Company of India, 1980); and Barry Bishop, *Karnali Zone under Stress* (Chicago: University of Chicago Press, 1990).

23. For an overview of energy problems in the subsistence economies of the Himalaya, see Pradeep Monga and P. Venkata Raman (eds.), *Energy, Environment, and Sustainable Development in the Himalayas* (New Delhi: Indus Publishing Co., 1992).

24. For example, see the discussion of forest-products usage among tribal groups in eastern Nepal in Ephrosine Danigellis, "*Jangal* Resource Use: Adaptive Strategies of Rais and Sherpas in the Upper Arun Valley of Eastern Nepal," in Michael Allen (ed.), *Anthropology of Nepal* (Kathmandu: Mandala Book Point, 1994).

CHAPTER 11: POVERTY AND ENVIRONMENTAL RISK

1. The reported figures are calculated from data obtained in the following sources: United Nations Development Program, *Human Development Report 1995* (New York and Oxford: Oxford University Press, 1995); International Monetary Fund, *International Financial Statistics Yearbook 1995* (Washington, D.C.: IMF); World Bank, *World Tables 1995* (Baltimore: Johns Hopkins University Press, 1995); A. N. Agrawal et al., *India Economic Information Yearbook 1995* (New Delhi: National Publishing House, 1995).

2. For a careful assessment of the primary indicators of poverty and levels of economic development in the Himalaya, see David C. Pitt, "Crisis, Pseudocrisis, or Supercrisis: Poverty, Women and Young People in the Himalaya: A Survey of Recent Developments," *Mountain Research and Development* 6, no. 2 (1986): 119–31. See also *Nepal: Poverty and Incomes* (Washington, D.C.: World Bank, 1991).

3. An insightful analysis of environmental resistance struggles in the Himalaya is

found in Ramachandra Guha, *The Unquiet Woods: Ecological Change and Peasant Resistance in the Himalaya* (Berkeley: University of California Press, 1990).

4. For a discussion of this issue in the case of the western Himalayan regions, see Haripriya Rangan, "Romancing the Environment: Popular Environmental Action in the Garhwal Himalaya," in John Friedman and Haripriya Rangan (eds.), *In Defence of Livelihood: Comparative Studies on Environmental Action* (West Hartford: Kumerian Press, 1993), 155–81. For a broader treatment of the subject, see John Isbister, *Promises Not Kept: Betrayal of Social Change in the Third World* (West Hartford, Conn.: Kumerian Publishers, 1993); Edward Said, *Culture and Imperialism* (New York: A. A. Knopf, 1993).

5. For a detailed analysis of women's work load in Nepal, see Meena Acharya and Lynn Bennett, *The Rural Women of Nepal* (Kathmandu: CEDA, 1980).

6. For discussions of this issue from women's perspectives, see Sumitra Gurung, "Gender Dimensions of Eco-Crisis and Resource Management in Nepal," In Michael Allen (ed.), *Anthropology of Nepal* (Kathmandu: Mandala Book Point, 1994); Maria Mies, Bina Pradhan, and Katherine Rankin, *Perspectives on the Role of Women in Mountain Development: Two Papers* (Kathmandu: ICIMOD discussion paper series MPE no. 1, 1990).

7. N. S. Jodha, *Perspectives on Poverty-Generating Processes in Mountain Areas* (Kathmandu: ICIMOD discussion paper series MFS no. 34, 1993).

8. For an overview of tourism development in the Himalaya, see *Mountain Tourism in Himachal Pradesh and the Hill Districts of Uttar Pradesh 1995* (Kathmandu:ICIMOD discussion paper series MEI no. 95/6, 1995); *Mountain Tourism in Nepal 1995* (Kathmandu: ICIMOD discussion paper series MEI no. 95/7, 1995). For specific reference to basic needs, see John Hough, "Bottom Up Versus Basic Needs: Integrating Conservation and Development in the Annapurna and Michiru Mountains of Nepal and Malawi," *Ambio* 18, no. 8 (1989): 434–41. For a critical perspective on tourism in the Himalaya, see David Zurick, "Adventure Travel and Sustainable Tourism in the Peripheral Economy of Nepal," *Annals of the Association of American Geographers* 82, no. 4 (1992):608–28; and David Zurick, *Errant Journeys: Adventure Travel in a Modern Age* (Austin: University of Texas Press, 1995).

CHAPTER 12: THE PALE OF MODERNITY

1. The agricultural output statistics reported here are determined from the following sources: Planning Commission, Royal Government of Bhutan, *Seventh Five-Year Plan* (Thimpu, 1991); Department of Agriculture, Himachal Pradesh, *A Note on Agricultural Development in Himachal Pradesh* (Shimla: Directorate of Agriculture, 1993); P. P. Karan, Hiroshi Ishii et al., *Nepal: Development and Change in a Landlocked Himalayan Kingdom* (Tokyo: Tokyo University of Foreign Studies, Monumenta Serindica no. 25, 1994); S. Mahendra Dev, *Indicators of Agricultural Unsustainability in the Indian Himalayas: A Survey* (Kathmandu: ICIMOD discussion paper series MFS no. 37, 1995); and World Bank, *Natural Resource Management for Sustainable Development: Nepal Study* (Kathmandu: HMG/N and World Bank, 1988).

2. For an excellent analysis of the circumstances of agricultural decline, see N. S. Jodha and S. Shrestha, "Sustainable and More Productive Mountain Agriculture: Problems and Prospects," in *Proceedings of the Tenth Anniversary Symposium of the In-*

ternational Center for Integrated Mountain Development, Dec 1–2, 1993 (Kathmandu: ICIMOD, 1994).

3. For studies that support this perspective, see, for example, Mahesh Banskota, *Agriculture and the Wider Market: Transformation Processes and Experiences of Bagmati Zone in Nepal* (Kathmandu: ICIMOD Occasional Paper no. 10, 1989); Bhimendra Katwal and Laxman Sah, *Transformation of Mountain Agriculture* (Kathmandu: ICIMOD discussion paper series MFS no. 26, 1992).

4. *Agricultural Development Experiences in Himachal Pradesh, India* (Kathmandu: ICIMOD MFS Workshop in Manali, India, report no. 1, 1988). For discussions of traditional irrigation and the government-sponsored expansion of irrigation in contrasting Himalaya localities, see Nigel J. R. Allan, "Household Food Supply in Hunza Valley, Pakistan," *Geographical Review* 80, no. 4 (1990):400–15; Narayan Dhakal et al., "Farmer-Managed and Government-Managed Irrigation Systems in Nepal," *Journal of the Faculty of Agriculture, Hokkaido University* 66, no. 1 (1994):139–50.

5. James Scott, *The Moral Economy of the Peasant* (Princeton: Princeton University Press, 1975).

6. Herman Kreutzmann, "Globalization, Spatial Integration, and Sustainable Development in Northern Pakistan," *Mountain Research and Development* 15, no. 3 (1995):213–27.

7. Stacy Pigg, "Unintended Consequences: The Ideological Impact of Development in Nepal," *South Asia Bulletin* 13, no. 1/2 (1993):45–58.

8. Stanley Stevens and Mingma Sherpa, "Indigenous Peoples and Protected Areas: New Approaches to Conservation in Highland Nepal," in Lawrence Hamilton et al. (eds.), *Parks, Peaks, and People* (Honolulu: East-West Center, 1993), 73–88; *The Makalu-Barun National Park and Conservation Area Management Plan* (Kathmandu: Department of Parks and Wildlife Conservation, HMG and Woodlands Mountain Institute, 1990); David Zurick, "The Road to Shangri La Is Paved: Spatial Development and Rural Transformation in Nepal," *South Asia Bulletin* 13, no. 1/2 (1993):35–44.

9. Mingma Sherpa, Sangay Wangchuk, and Til Bahadur Mongar, "Designing a Protected Area System in the Himalaya: The Bhutan Approach," in Hamilton et al., *Parks, Peaks, and People*, 174–87.

10. For an evocative portrayal of the salt trade in Nepal, see Eric Valli and Diane Sommers, *Caravans of the Himalaya* (Washington, D.C.: National Geographic Society, 1995).

11. Nigel J. R. Allan, "Impact of Afghan Refugees on the Vegetation Resources of Pakistan's Hindu Kush-Himalaya," *Mountain Research and Development* 7, no. 2 (1987):200–204.

12. The statistics on urban development are drawn from the following sources: Central Bureau of Statistics, Nepal, *Nepal 1991 Census, Preliminary Result* (Kathmandu; Pakistan Bureau of Statistics, 1991); *Census Report of Pakistan, 1991* (Islamabad); Royal Government of Bhutan, Planning Commission, *Seventh Five-Year Plan (1992–1997)* (Thimpu, 1991); Pitamber Sharma, *Urbanization in Nepal*, Papers of the East-West Population Institute no. 110 (Honolulu: East-West Center, 1989); Pitamber Sharma, "Population and Employment Challenges in the Mountains," *Proceedings of the Tenth Anniversary Symposium of ICIMOD* (Kathmandu: ICIMOD, 1994).

13. An overview of Nepal's planned urban growth is provided in PADCO, *Nepal: Urban Development Assessment* (Kathmandu: HMG/USAID, 1984).

14. For a discussion of the Karakoram Highway in Pakistan and road building in Ladakh, see Lewis Owen, "High Roads, High Risks," *The Geographical Magazine* 67, no. 1 (1996):12–15.

15. See for example, Nigel Allan, 1986, "Accessibility and Altitudinal Zonation Models of Mountains," *Mountain Research and Development* 6, no. 3 (1986):185–94; and Nigel Allan, "Highways to the Sky," *Tourism Recreation Research* 13(1986):11–16.

16. B. B. Deoja, "Planning and Management of Mountain Roads and Infrastructures," *Proceedings of the International Symposium on Mountain Environment and Development* (Kathmandu: ICIMOD, 1994), 4.

17. Quoted in Mahesh Banskota and Archana Karki (eds.), *Sustainable Development of Fragile Mountain Areas of Asia* (Kathmandu: ICIMOD Regional Conference Report, 1994), 7.

CHAPTER 13: THE CONTROL OF NATURE

1. *Master Forestry Plan for the Forestry Sector Nepal* (Kathmandu: HMGN/ADB/ FINNIDA, Ministry of Forests and Soil Conservation, 1988); Water and Energy Commission Secretariat, *Land Use in Nepal: A Summary of the Land Resources Mapping Project Results (with Special Emphasis on Forest Land Use)* (Kathmandu: WECS, 1986).

2. Department of Forest, Farming, and Conservation, *Forests of Himachal Pradesh* (Shimla, 1993).

3. *Bhutan: Towards Sustainable Development in a Unique Environment*, National Report for the United Nations Conference on Environment and Development (Thimphu, n.d.); Food and Agriculture Organization, *Master Plan for Forestry Development in Bhutan, Wood Energy Sector Analysis* (Thimpu: Ministry of Agriculture, Department of Forestry [FAO field document 32, GCP/RAS/131/NET], 1991).

4. Sunderlal Bahugna, "Chipko: The Peoples' Movement to Protect Forests," *Cultural Survival Quarterly* 19, no. 3 (1987):27–30. For a review of environmental activism in the Garhwal region, especially in the Doon Valley, see Prem Thadhani, *Chronicles of the Doon Valley* (New Delhi: Indus Publishing Co., 1993).

5. Department of Horticulture, *A Glimpse of Himachal Horticulture—Facts in Figures* (Shimla, n.d.).

6. Agricultural Projects Services Center, *Nepal Hillfruit Development Project, Baseline Survey* (Kathmandu: APROSC, 1991); Brian Carson, *Master Plan for Horticulture Development* (Kathmandu and Manila: HMG, Ministry of Agriculture and ADB, 1990).

7. The data on Kotgarh orchards were obtained during field studies in the region in 1994. Interviews were conducted with orchardists, sawmill operators, apple traders, workers, and government officials. For an overview of apple growing in the Himalaya, see S. S. Teaotia, *Horticulture Development in the Hindu Kush-Himalaya Region* (New Delhi: Oxford University Press and IBH Publishing Co, 1993).

8. The data on food consumption and deficits were obtained from surveys of village farmers in Kotgarh and interviews with agriculture extension workers in the region.

9. Pre-harvest contracts are generally made between the orchard owners and entrepreneurs based in the plains market towns. The contractors bear the risk of poor harvests and the cost of transportation. In return, they receive favorable prices for apples. Under these arrangements, the growers avoid some of the risks of apple

growing. But most of the profit goes to the contractors who have the market connections. For additional discussion, see also R. Swarup, *Agricultural Economy of Himalaya Region* (Nainital: G. B. Pant Institute of Himalayan Environment and Development, 1991).

10. International Center for Integrated Mountain Development, *Sustainable Development of Fragile Mountain Areas of Asia.* Regional Conference Report, December 13–15, 1994 (Kathmandu: ICIMOD, 1995).

11. L. A. Bruijnzeel and C. N. Bremer, *Highland-Lowland Interactions in the Ganges Brahmaputra River Basin* (Kathmandu: ICIMOD Occasional Paper no. 11, 1989), 42.

12. For an analysis of Himalaya river discharges, see Donald Alford, *Hydrological Aspects of the Himalayan Region* (Kathmandu: ICIMOD Occasional Paper no. 18, 1992); also, B. C. Upreti, *Politics of Himalayan River Waters* (New Delhi: Nirala Publications, 1993).

13. *Jawaharlal Nehru's Speech at the Silver Jubilee Celebrations,* CBIP, November 17, 1952; quoted in Upreti, *Politics of Himalayan River Waters,* 60.

14. Indian National Trust for Art and Cultural Heritage, *The Tehri Dam: A Prescription for Disaster* (New Delhi: INTACH, 1987); Jayanta Bandyopadhyay and Dipak Gyawali, 1994, "Himalayan Water Resources: Ecological and Political Aspects of Management," *Mountain Research and Development* 14, no. 1 (1987):1–24.

15. Vijay Paranjpye, *Evaluating the Tehri Dam* (New Delhi: Indian National Trust for Art and Cultural Heritage, 1988).

16. Ajaya Dixit, "Mapping Nepal's Water Resource," *Himal* 8, no. 4 (1995):23. For an overview of hydropower in Nepal, see D. B. Thapa, "Geological Study in Hydropower Development in Nepal," *Bulletin, Department of Geology, Tribhuvan University* 3, no. 1 (1993):35–41. For a critique of the Kosi River high dam project, see Rajendra Daha, "Gunning for Kosi High," *Himal* 8, no. 4 (1995):24–28.

CHAPTER 14: A MOUNTAIN THEME PARK?

1. For an overview of adventure travel and its impacts on remote world places, see David Zurick, *Errant Journeys: Adventure Travel in a Modern Age* (Austin: University of Texas Press, 1995).

2. P. P. Karan, Hiroshi Ishii et al., *Nepal: Development and Change in a Landlocked Himalayan Kingdom* (Tokyo: Tokyo University of Foreign Studies, Monumenta Serindica no. 25, 1994); see especially chapter 10, "Development of Tourism," 215–34.

3. Planning Commission, Royal Government of Bhutan, *Seventh Five-Year Plan (1992–1997)* (Thimpu, 1991).

4. International Center for Integrated Mountain Development, *Mountain Tourism in Himachal Pradesh and the Hill Districts of Uttar Pradesh* (Kathmandu: ICIMOD Discussion Paper Series no. MEI 95/6, 1995). See also P. K. Chakravarti, "Tourist Industry in Darjeeling," *Himalayan Research and Development* 1, no. 2 (1982):177–80; International Center for Integrated Mountain Development, *Annual Report 1995* (Kathmandu: ICIMOD, 1995), 28.

5. David Greenwood, "Culture by the Pound: An Anthropological Perspective on Tourism as Cultural Commoditization," in Valerie Smith (ed.), *Hosts and Guests: The Anthropology of Tourism* (Philadelphia: University of Pennsylvania Press, 1989), 171–86, 179.

6. For a detailed look at Sherpa culture in relation to tourism, see Vincanne Adams, *Virtual Sherpas* (Princeton: Princeton University Press, 1996).

7. See, for example, Stanley Stevens, *Claiming the High Ground: Sherpas, Subsistence, and Environmental Change in the Highest Himalaya* (Berkeley: University of California Press, 1993); James Fisher, *The Sherpas: Reflections on Change in Himalayan Nepal* (Berkeley: University of California Press, 1990); David Zurick, "Adventure Travel and Sustainable Tourism in the Peripheral Economy of Nepal," *Annals of the Association of American Geographers* 82, no. 4 (1992):608–28; International Center for Integrated Mountain Development, *Mountain Tourism in Nepal* (Kathmandu: ICIMOD discussion paper series MEI no. 95/7, 1995).

8. ICIMOD, *Mountain Tourism in Nepal.*

9. For a comprehensive overview of protected areas in the Himalaya region, see Michael Green, *Nature Reserves of the Himalaya and the Mountains of Central Asia* (Oxford: Oxford University Press and IUCN—The World Conservation Union, 1993).

10. There is a large and growing literature on ecotourism; see, for example, Elizabeth Boo, *Ecotourism: The Potentials and Pitfalls*, vols. 1 and 2 (Washington, D.C.: World Wildlife Fund, 1990). For a critique of ecotourism, see Zurick, *Errant Journeys.*

11. Stanley Stevens and Mingma Sherpa, "Indigenous Peoples and Protected Areas: New Approaches to Conservation in Highland Nepal," in Lawrence Hamilton et al. (eds.), *Parks, Peaks, and People* (Honolulu: East-West Center, 1993), 73–88.

CHAPTER 15: LANDSCAPES OF THE FUTURE

1. The Nepalese geographer Harka Gurung made a similar point in a recent address to the International Center for Integrated Mountain Development in Kathmandu: Harka Gurung, "Conservation of the Environment in the Nepal Himalayas," *ICIMOD Newsletter* 24 (1996):2.

2. Christopher Flavin, "Sustainable Development in the Kingdom of Bhutan," discussion paper for the National Workshop on Sustainable Development and the Environment, May 4–5, 1990 (Paro).

3. *Bhutan: Towards Sustainable Development in a Unique Environment*, National Report for the United Nations Conference on Environment and Development (Thimphu, n.d.), 40–41.

4. World Bank, 1988, *Natural Resource Management for Sustainable Development: A Study of Feasible Policies, Institutions, and Investment Activities in Nepal with Special Emphasis on the Hills* (Kathmandu: HMG/World Bank/ERL, 1988), 142–55.

5. For a specific village study, see Jefferson Fox, "Forest Resources in a Nepali Village in 1980 and 1990: The Positive Influence of Population Growth," *Mountain Research and Development* 13, no. 1 (1994):89–98. The national trends are discussed in part 3 of the book and are based on extensive field study; on the analysis of vegetation maps produced by remote sensing analysis (cf. M. A. Kawosa, *Remote Sensing of the Himalaya* [Dehra Dun: Natraj Publishers, 1988]); and on the interpretation of study results of the Land Resources Mapping Project (LRMP) reported in WEC (Water and Energy Commission), *Land Use in Nepal* (Kathmandu: Ministry of Water Resources Report no. 4/1/310386/1/1, 1986).

6. Central Bureau of Statistics, *Statistical Yearbook of Nepal* (Kathmandu, 1993); Planning and Development Consulting Organization, *Nepal Urban Development Assessment* (Kathmandu: HMG/USAID, 1984).

7. HMG/Environment Protection Council, *Nepal Environmental Policy and Action Plan* (Kathmandu, 1993), xi.

8. International Center for Integrated Mountain Development, *International Symposium on Mountain Environment and Development: Constraints and Opportunities* (Kathmandu: ICIMOD, 1994), 22–23. The regional initiatives help to inform government policy and planning programs, articulated in the various annual plans of the Himalaya countries and regions. See, for example, Government of Arunachal Pradesh, *Annual Plan 1995–96* (Itanagar); Government of Sikkim, *Draft Eighth Five-Year Plan, 1992–97* (Gangtok).

9. Government of Pakistan, *National Report Submitted to the United Nations Conference on Environment and Development* (Islamabad: Environment and Urban Affairs Division, 1991).

10. Derek Denniston, *High Priorities: Conserving Mountain Ecosystems and Cultures* (Washington, D.C.: Worldwatch Institute paper no. 123, 1995); G. B. Pant Institute of Himalayan Environment and Development, *Action Plan for the Himalaya,* Himavikas Occasional Paper no. 2 (Almora, 1992); United Nations Commission on Environment and Development, *Agenda 21* (New York: United Nations, 1992); HMG, *The National Conservation Strategy for Nepal* (Kathmandu: HMG and IUCN, 1988); Mahesh Banskota, *Integrated Planning for Mountain Environment and Development.* International Symposium on Mountain Environment and Development, Dec. 1–3, 1993 (Kathmandu: ICIMOD, 1994).

11. Nanda Shrestha, *Landlessness and Migration in Nepal* (Boulder: Westview Press, 1990).

12. See, for example, Kirsten Johnson, Elizabeth Olson, and Sumitra Manandhar, "Environmental Knowledge and Response to Natural Hazard in Mountainous Nepal," *Mountain Research and Development* 2, no. 2 (1982):175–88; David Zurick, "Traditional Knowledge and Conservation as a Basis for Development in a West Nepal Village," *Mountain Research and Development* 10, no. 1 (1990):23–33.

13. Alan Durning, *Guardians of the Land: Indigenous People and the Wealth of the Earth* (Washington, D.C.: Worldwatch Institute paper no. 112, 1992); *Indigenous Knowledge Systems and Biodiversity Management,* proceedings of a MacArthur Foundation ICIMOD Seminar, April 13–14, 1994 (Kathmandu).

14. "Uttarkhand Movement: A Message Loud and Clear" *Indian Express,* October 17, 1994, 7.

15. U. Dhar, *Himalayan Biodiversity: Conservation Strategies,* G. B. Pant Institute of Himalayan Environment and Development, Himavikas Occasional Paper no. 3 (Almora, 1993).

Index

postcolonial geopolitics, 64

postcolonial nationalism, 49–50

poverty: conditions of, 193–95; and cultural survival, 49; and ecological crisis, 49; and environmental degradation, 191, 282; and land degradation, 135; roots of, 193; and women, 196–97

Pradhan, Bina, 340

Prakash, Sanjeev, 333

Price, Martin, 336

Prithvinarayan Shah, 82

Pudup, Mary Beth, 330

Raha, Manis K., 334

Rahul, Ram, 332

Rais, 75

Raman, P. Venkata, 339

Rana: autocracy, 94, 107; family, 90, 106; oligarchy, 112

Rangan, Haripriya, 333, 340

rangeland management, 74

Ranjit Singh, 72

Rankin, Katherine, 340

Rapti Integrated Rural Development Project, 210

Redclift, Michael, 328

regional analysis, x; of century-old data, xii

regional development strategy, 38

regional goals: at odds with local goals, 44; political and administrative, 118

regional integration, 210

regional markets, 198

Regmi, Mahesh C., 331, 333

resource management: blends of native ways and institutional guidelines in, 17–18; state intervention in, 92; traditional forms of, 16

revenue: concentration in cities and towns, 45; during Gorkha rule, 86–87; during Rana family rule, 90

Richard, J. F., 331, 333

Richu Khola, Sikkim, 154; agriculture and population in, 155; land management practices, 158–59

roads: and development pattern, 211; and environmental damage, 219; expansion of, 209; and forest exploitation, 111; in Nepal, 118, 217–18; newly built in

Bhutan, Sikkim, and Nepal, 42, 218; social costs of, 43; in transformation of society and land use, 53, 54; in Western Himalaya, 117, 216

Rose, Leo, 331–32

Roy, B. K., xi

Roy, N. R., 332–33

Royal Chitawan National Park, 262, 264

Royal Geographical Society, 95, 102

Royal Manas National Park, 36, 211, 263, 283, 294

Rupi Bhabha Sanctuary, 262

Russia: advances into the mountains, 102; interests in land north of the Himalaya, 107

Sagarmatha-Qomolongma Park (Sagarmatha National Park), 260, 262, 265–67, 294

Sah, Laxman, 341

Said, Edward, 340

Sanjay Vidyut Hydroelectric Project, 262

Sargent, Caroline, 338

Sargent, Orlando, 338

Sayer, A., 330

Schikoff, U., 338

Scholz, John, 332

Schroder, John F., 325

Schweinfurth, U., 328

Scott, James, 341

sedentary farming, 52

Segauli Treaty of 1815, 91, 102

separatist struggles, 50, 121–22

Shah Kingdom, 84, 89

Sharma, Pitamber, 341

Shepherd, K. R., 331, 338

Sherdukpen, 69

Sherpa, 212, 257, 266–67

Sherpa, Mingma, 341

shifting cultivation, 74, 90, 110, 174, 182, 206

Shimla, *100*; pollution in, 246

Shimla Ecology Project, 262, 283

Shimla Water Catchment Sanctuary, 262

Shrestha, C. L., 326

Shrestha, Nanda, 338, 345

Shrestha, R. L., 335

Shrestha, S., 337, 340

Shrestha, S. L., 326

village: competing economic and environmental conflicts, 49; connection with wider world, 47, 57; and design and management of economic programs, 224; farmland requirement, 189; ideals, 190; and resource management, 266; and sustainable development, 288; use of forest products, 189

Vuichard, D., 336

Wangchuk, Sangay, 341
Wangchuk, Ugyen, 98, 112
Warburton, Thomas, 326, 334
Water Research Institute, India, 141

Watts, Michael, 328
wealth: and control of resources, 48; and land ownership, 196
women: activism by, 197; burden on, 196; environmental knowledge held by, 289
Woodlands Mountain Institute, 211

Yandaboo Treaty, 95

Zimmermann, M., 336
Ziro Valley, *50, 176, 177*
Zurick, David, 325, 330–31, 334, 337, 340–41, 343–45, 347

DAVID ZURICK was raised along the shores of Lake Huron and obtained his undergraduate and master's degrees in geography from Michigan State University. He completed his Ph.D. in geography in 1986 at the University of Hawaii. Currently he is a professor of geography at Eastern Kentucky University and lives in a tributary valley among the foothills of the Cumberland Plateau near Berea, Kentucky.

Professor Zurick has conducted geographic field studies and photographic surveys throughout the Himalaya since 1975. His fieldwork has been supported by grants and awards from the National Science Foundation, the East-West Center, and the Kentucky Council for International Education. He was nominated a Fellow of the Explorers Club in 1987 and has held various memberships in scientific and humanistic organizations.

David Zurick is the author of two previous books, *Errant Journeys* and *Hawaii Naturally*.

P. P. KARAN is a professor of geography at the University of Kentucky and chair of the university's Japan Studies Committee. Born in India, Professor Karan has lived in the United States since 1954. His sustained research contributions on the Himalaya and Tibet have been recognized through awards and grants from the Population

Council, the American Philosophical Society, the National Geographic Society, the American Geographical Society, and the National Science Foundation. Among other awards, he has received the Sir George Everest Gold Medal for cartographic and resource management research in the Himalaya. He is a Fellow of the Explorers Club.

Professor Karan has written fifteen books, contributed essays to twenty-five books, and co-edited or co-authored four books. His most recent books are *Nepal: A Himalayan Kingdom in Transition* (co-authored with Hiroshi Ishii), *The Japanese City*, and *The Japanese Landscapes* (co-authored with Cotton Mather and Shigeru Iijima).

LIBRARY OF CONGRESS CATALOGING-IN-PUBLICATION DATA

Zurick, David
    Himalaya : life on the edge of the world / by David Zurick and P. P. Karan ;
maps by Julsun Pacheco.
        p.   cm.
    Includes bibliographical references and index.
    ISBN 0-8018-6168-3 (alk. paper)
    1. Himalaya Mountains Region.   I. Karan, Pradyumna P. (Pradyumna Prasad).
II. Pacheco, Julsun.
DS485.H6Z87   1999
954.96—dc21   99-11037
                                                          CIP